MONOGRAPHS ON STATISTICS AND APPLIED PROBABILITY

General Editors

V. Isham, N. Keiding, T. Louis, N. Reid, R. Tibshirani, and H. Tong

1 Stochastic Population Models in Ecology and Epidemiology *M.S. Barlett* (1960)
2 Queues *D.R. Cox and W.L. Smith* (1961)
3 Monte Carlo Methods *J.M. Hammersley and D.C. Handscomb* (1964)
4 The Statistical Analysis of Series of Events *D.R. Cox and P.A.W. Lewis* (1966)
5 Population Genetics *W.J. Ewens* (1969)
6 Probability, Statistics and Time *M.S. Barlett* (1975)
7 Statistical Inference *S.D. Silvey* (1975)
8 The Analysis of Contingency Tables *B.S. Everitt* (1977)
9 Multivariate Analysis in Behavioural Research *A.E. Maxwell* (1977)
10 Stochastic Abundance Models *S. Engen* (1978)
11 Some Basic Theory for Statistical Inference *E.J.G. Pitman* (1979)
12 Point Processes *D.R. Cox and V. Isham* (1980)
13 Identification of Outliers *D.M. Hawkins* (1980)
14 Optimal Design *S.D. Silvey* (1980)
15 Finite Mixture Distributions *B.S. Everitt and D.J. Hand* (1981)
16 Classification *A.D. Gordon* (1981)
17 Distribution-Free Statistical Methods, 2nd edition *J.S. Maritz* (1995)
18 Residuals and Influence in Regression *R.D. Cook and S. Weisberg* (1982)
19 Applications of Queueing Theory, 2nd edition *G.F. Newell* (1982)
20 Risk Theory, 3rd edition *R.E. Beard, T. Pentikäinen and E. Pesonen* (1984)
21 Analysis of Survival Data *D.R. Cox and D. Oakes* (1984)
22 An Introduction to Latent Variable Models *B.S. Everitt* (1984)
23 Bandit Problems *D.A. Berry and B. Fristedt* (1985)
24 Stochastic Modelling and Control *M.H.A. Davis and R. Vinter* (1985)
25 The Statistical Analysis of Composition Data *J. Aitchison* (1986)
26 Density Estimation for Statistics and Data Analysis *B.W. Silverman* (1986)
27 Regression Analysis with Applications *G.B. Wetherill* (1986)
28 Sequential Methods in Statistics, 3rd edition
G.B. Wetherill and K.D. Glazebrook (1986)
29 Tensor Methods in Statistics *P. McCullagh* (1987)
30 Transformation and Weighting in Regression
R.J. Carroll and D. Ruppert (1988)
31 Asymptotic Techniques for Use in Statistics
O.E. Bandorff-Nielsen and D.R. Cox (1989)
32 Analysis of Binary Data, 2nd edition *D.R. Cox and E.J. Snell* (1989)
33 Analysis of Infectious Disease Data *N.G. Becker* (1989)
34 Design and Analysis of Cross-Over Trials *B. Jones and M.G. Kenward* (1989)

35 Empirical Bayes Methods, 2nd edition *J.S. Maritz and T. Lwin* (1989)
36 Symmetric Multivariate and Related Distributions
   K.T. Fang, S. Kotz and K.W. Ng (1990)
37 Generalized Linear Models, 2nd edition *P. McCullagh and J.A. Nelder* (1989)
38 Cyclic and Computer Generated Designs, 2nd edition
   J.A. John and E.R. Williams (1995)
39 Analog Estimation Methods in Econometrics *C.F. Manski* (1988)
40 Subset Selection in Regression *A.J. Miller* (1990)
41 Analysis of Repeated Measures *M.J. Crowder and D.J. Hand* (1990)
42 Statistical Reasoning with Imprecise Probabilities *P. Walley* (1991)
43 Generalized Additive Models *T.J. Hastie and R.J. Tibshirani* (1990)
44 Inspection Errors for Attributes in Quality Control
   N.L. Johnson, S. Kotz and X. Wu (1991)
45 The Analysis of Contingency Tables, 2nd edition *B.S. Everitt* (1992)
46 The Analysis of Quantal Response Data *B.J.T. Morgan* (1992)
47 Longitudinal Data with Serial Correlation—A state-space approach
   R.H. Jones (1993)
48 Differential Geometry and Statistics *M.K. Murray and J.W. Rice* (1993)
49 Markov Models and Optimization *M.H.A. Davis* (1993)
50 Networks and Chaos—Statistical and probabilistic aspects
   O.E. Barndorff-Nielsen, J.L. Jensen and W.S. Kendall (1993)
51 Number-Theoretic Methods in Statistics *K.-T. Fang and Y. Wang* (1994)
52 Inference and Asymptotics *O.E. Barndorff-Nielsen and D.R. Cox* (1994)
53 Practical Risk Theory for Actuaries
   C.D. Daykin, T. Pentikäinen and M. Pesonen (1994)
54 Biplots *J.C. Gower and D.J. Hand* (1996)
55 Predictive Inference—An introduction *S. Geisser* (1993)
56 Model-Free Curve Estimation *M.E. Tarter and M.D. Lock* (1993)
57 An Introduction to the Bootstrap *B. Efron and R.J. Tibshirani* (1993)
58 Nonparametric Regression and Generalized Linear Models
   P.J. Green and B.W. Silverman (1994)
59 Multidimensional Scaling *T.F. Cox and M.A.A. Cox* (1994)
60 Kernel Smoothing *M.P. Wand and M.C. Jones* (1995)
61 Statistics for Long Memory Processes *J. Beran* (1995)
62 Nonlinear Models for Repeated Measurement Data
   M. Davidian and D.M. Giltinan (1995)
63 Measurement Error in Nonlinear Models
   R.J. Carroll, D. Rupert and L.A. Stefanski (1995)
64 Analyzing and Modeling Rank Data *J.J. Marden* (1995)
65 Time Series Models—In econometrics, finance and other fields
   D.R. Cox, D.V. Hinkley and O.E. Barndorff-Nielsen (1996)
66 Local Polynomial Modeling and its Applications *J. Fan and I. Gijbels* (1996)
67 Multivariate Dependencies—Models, analysis and interpretation
   D.R. Cox and N. Wermuth (1996)

68 Statistical Inference—Based on the likelihood *A. Azzalini* (1996)
69 Bayes and Empirical Bayes Methods for Data Analysis
B.P. Carlin and T.A Louis (1996)
70 Hidden Markov and Other Models for Discrete-Valued Time Series
I.L. Macdonald and W. Zucchini (1997)
71 Statistical Evidence—A likelihood paradigm *R. Royall* (1997)
72 Analysis of Incomplete Multivariate Data *J.L. Schafer* (1997)
73 Multivariate Models and Dependence Concepts *H. Joe* (1997)
74 Theory of Sample Surveys *M.E. Thompson* (1997)
75 Retrial Queues *G. Falin and J.G.C. Templeton* (1997)
76 Theory of Dispersion Models *B. Jørgensen* (1997)
77 Mixed Poisson Processes *J. Grandell* (1997)
78 Variance Components Estimation—Mixed models, methodologies and applications
P.S.R.S. Rao (1997)
79 Bayesian Methods for Finite Population Sampling
G. Meeden and M. Ghosh (1997)
80 Stochastic Geometry—Likelihood and computation
O.E. Barndorff-Nielsen, W.S. Kendall and M.N.M. van Lieshout (1998)
81 Computer-Assisted Analysis of Mixtures and Applications—
Meta-analysis, disease mapping and others *D. Böhning* (1999)
82 Classification, 2nd edition *A.D. Gordon* (1999)
83 Semimartingales and their Statistical Inference *B.L.S. Prakasa Rao* (1999)
84 Statistical Aspects of BSE and vCJD—Models for Epidemics
C.A. Donnelly and N.M. Ferguson (1999)
85 Set-Indexed Martingales *G. Ivanoff and E. Merzbach* (2000)
86 The Theory of the Design of Experiments *D.R. Cox and N. Reid* (2000)
87 Complex Stochastic Systems
O.E. Barndorff-Nielsen, D.R. Cox and C. Klüppelberg (2001)
88 Multidimensional Scaling, 2nd edition *T.F. Cox and M.A.A. Cox* (2001)
89 Algebraic Statistics—Computational Commutative Algebra in Statistics
G. Pistone, E. Riccomagno and H.P. Wynn (2001)
90 Analysis of Time Series Structure—SSA and Related Techniques
N. Golyandina, V. Nekrutkin and A.A. Zhigljavsky (2001)
91 Subjective Probability Models for Lifetimes
Fabio Spizzichino (2001)
92 Empirical Likelihood *Art B. Owen (2001)*
93 Statistics in the 21st Century *Adrian E. Raftery, Martin A. Tanner,
and Martin T. Wells (2001)*
94 Accelerated Life Models: Modeling and Statistical Analysis
Vilijandas Bagdonavičius and Mikhail Nikulin (2001)
95 Subset Selection in Regression, Second Edition
Alan Miller (2002)
96 Topics in Modelling of Clustered Data
Marc Aerts, Helena Geys, Geert Molenberghs, and Louise M. Ryan (2002)
97 Components of Variance *D.R. Cox and P.J. Solomon* (2002)

Hidden Markov and Other Models for Discrete-valued Time Series

Iain L. MacDonald
*University of Cape Town
South Africa*

and

Walter Zucchini
*University of Göttingen
Germany*

CHAPMAN & HALL/CRC

Boca Raton   London   New York   Washington, D.C.

Library of Congress Cataloging-in-Publication Data

Catalog record is available from the Library of Congress

This book contains information obtained from authentic and highly regarded sources. Reprinted material is quoted with permission, and sources are indicated. A wide variety of references are listed. Reasonable efforts have been made to publish reliable data and information, but the author and the publisher cannot assume responsibility for the validity of all materials or for the consequences of their use.

Apart from any fair dealing for the purpose of research or private study, or criticism or review, as permitted under the UK Copyright Designs and Patents Act, 1988, this publication may not be reproduced, stored or transmitted, in any form or by any means, electronic or mechanical, including photocopying, microfilming, and recording, or by any information storage or retrieval system, without the prior permission in writing of the publishers, or in the case of reprographic reproduction only in accordance with the terms of the licenses issued by the Copyright Licensing Agency in the UK, or in accordance with the terms of the license issued by the appropriate Reproduction Rights Organization outside the UK.

The consent of CRC Press LLC does not extend to copying for general distribution, for promotion, for creating new works, or for resale. Specific permission must be obtained in writing from CRC Press LLC for such copying.

Direct all inquiries to CRC Press LLC, 2000 N.W. Corporate Blvd., Boca Raton, Florida 33431.

Trademark Notice: Product or corporate names may be trademarks or registered trademarks, and are used only for identification and explanation, without intent to infringe.

Visit the CRC Press Web site at www.crcpress.com

© 1997 by Iain L. MacDonald and Walter Zucchini
First edition 1997
First CRC Press reprint 2000
Originally published by Chapman & Hall

No claim to original U.S. Government works
International Standard Book Number 0-412-55850-5
Printed in the United States of America   2 3 4 5 6 7 8 9 0
Printed on acid-free paper

TO HEINZ LINHART

Contents

Preface ... xiii

PART ONE   Survey of models ... 1

1 **A survey of models for discrete-valued time series** ... 3
   1.1 Introduction: the need for discrete-valued time series models ... 3
   1.2 Markov chains ... 6
       1.2.1 Saturated Markov chains ... 6
       1.2.2 A nonhomogeneous Markov chain model for binary time series ... 11
   1.3 Higher-order Markov chains ... 12
   1.4 The DARMA models of Jacobs and Lewis ... 17
   1.5 Models based on thinning ... 21
       1.5.1 Models with geometric marginal ... 21
       1.5.2 Models with negative binomial marginal ... 23
       1.5.3 Models with Poisson marginal ... 25
       1.5.4 Models with binomial marginal ... 29
       1.5.5 Results not based on any explicit distributional assumption ... 30
   1.6 The bivariate geometric models of Block, Langberg and Stoffer ... 32
       1.6.1 Moving average models with bivariate geometric distribution ... 32
       1.6.2 Autoregressive and autoregressive moving average models with bivariate geometric distribution ... 34
   1.7 Markov regression models ... 37
   1.8 Parameter-driven models ... 42
   1.9 State-space models ... 45

x   CONTENTS

    1.10 Miscellaneous models   47
    1.11 Discussion   49

PART TWO   Hidden Markov models   **53**

2 The basic models   **55**
    2.1 Introduction   55
    2.2 Some theoretical aspects of hidden Markov models in speech processing   57
    2.3 Hidden Markov time series models: definition and notation   65
    2.4 Correlation properties   69
        2.4.1 The autocorrelation function of a Poisson-hidden Markov model   70
        2.4.2 The autocorrelation function of a binomial-hidden Markov model   74
        2.4.3 The partial autocorrelation function   77
    2.5 Evaluation of the likelihood function   77
    2.6 Distributional properties   80
        2.6.1 Marginal, joint and conditional distributions of the observations   80
        2.6.2 The Markov chain conditioned on the observations   84
        2.6.3 Runlength distributions for binary hidden Markov models   86
    2.7 Parameter estimation   90
        2.7.1 Computing maximum likelihood estimates   90
        2.7.2 Asymptotic properties of maximum likelihood estimators   95
        2.7.3 Use of the parametric bootstrap   96
    2.8 Identification of outliers   97
    2.9 Reversibility   101
    2.10 Discussion   105

3 Extensions and modifications   **109**
    3.1 Introduction   109
    3.2 Models based on a second-order Markov chain   110
    3.3 Multinomial-hidden Markov models   115
        3.3.1 The likelihood   116
        3.3.2 Marginal properties and cross-correlations   117
        3.3.3 A model for categorical time series   119
    3.4 Multivariate models   121

CONTENTS   xi

        3.4.1   The likelihood function for multivariate
                models   122
        3.4.2   Cross-correlations of models assuming con-
                temporaneous conditional independence   122
        3.4.3   Cross-correlations of models not assuming
                contemporaneous conditional independence   124
        3.4.4   Multivariate models with time lags   125
        3.4.5   Multivariate models in which some variables
                are discrete and others continuous   126
  3.5 Models with state-dependent probabilities depend-
       ing on covariates   128
  3.6 Models in which the Markov chain is homogeneous
       but not assumed stationary   129
  3.7 Models in which the Markov chain is nonhomoge-
       neous   130
  3.8 Joint models for the numbers of trials and the
       numbers of successes in those trials   133
  3.9 Discussion   135

4 Applications   **137**
  4.1 Introduction   137
  4.2 The durations of successive eruptions of the Old
       Faithful geyser   138
        4.2.1   Markov chain models   138
        4.2.2   Hidden Markov models   140
        4.2.3   Comparison of models   144
        4.2.4   Forecast distributions   146
  4.3 Epileptic seizure counts   146
  4.4 Births at Edendale hospital   152
        4.4.1   Models for the proportion Caesarean   152
        4.4.2   Models for the total number of deliveries   159
        4.4.3   Conclusion   161
  4.5 Locomotory behaviour of *Locusta migratoria*   162
        4.5.1   Multivariate models   163
        4.5.2   Univariate models   166
        4.5.3   Conclusion   167
  4.6 Wind direction at Koeberg   168
        4.6.1   Three hidden Markov models for hourly
                averages of wind direction   168
        4.6.2   Model comparisons and other possible models   173
        4.6.3   Conclusion   176
  4.7 Evapotranspiration   177

4.8	Thinly traded shares on the Johannesburg Stock Exchange	178
	4.8.1 Univariate models	179
	4.8.2 Multivariate models	180
	4.8.3 Discussion	182
4.9	Daily rainfall at Durban	184
4.10	Homicides and suicides, Cape Town, 1986–1991	191
	4.10.1 Models for firearm homicides as a proportion of all homicides, suicides and legal intervention homicides	191
	4.10.2 Models for the number of firearm homicides	194
	4.10.3 Firearm homicides as a proportion of all homicides, and firearm suicides as a proportion of all suicides	195
	4.10.4 Models for the proportions in each of the five categories of death	200
4.11	Conclusion	201

Appendices   **203**

A   Proofs of results used in the derivation of the Baum–Welch algorithm   203

B   Data   207

References   **217**

Author index   **227**

Subject index   **231**

Preface

Discrete-valued time series are common in practice, yet methods for their analysis are not well known. Many university courses on time series pay little or no attention to methods and models specifically designed for discrete-valued series, although the list of discrete-valued models proposed in the research literature is by now a long one. These models have not found their way into the major statistical packages nor, we believe it is fair to say, into the repertoire of most applied statisticians. The reason may be that there is no well-known family of models that are structurally simple, sufficiently versatile to cater for a useful variety of data types, and readily accessible to the practitioner.

We have two main objectives in this monograph. Firstly, we wish to provide a summary of the models that have been proposed and of such data-analytic methodology as has been developed for these. Secondly, we wish to describe in detail the class of hidden Markov models. In our opinion this class of models possesses enough of the desirable properties mentioned above to make it worthwhile for the applied statistician to become familiar with them. The two parts of the book reflect these two separate objectives, and can be read independently.

Our intended readership is primarily applied statisticians who meet discrete-valued time series in their work, and who need to know what appropriate methodology is available and how to apply it. The book is also intended for statistical researchers who wish to contribute to the further development of the subject. Finally, some of the material might profitably be included in a graduate — or even undergraduate — course on time series analysis.

Part One presents an account of the models (other than hidden Markov) that have been proposed. While we hope that our survey is reasonably comprehensive, this is certainly not a self-contained exposition of the models and their properties, and even falls short of what can properly be described as a review. What is offered is an annotated guide to the literature, in which the models have been classified in a way that we hope will be useful for the purposes of comparison and reference. However, the tangle of relationships that exist among models developed from different starting-points has defeated our attempts to provide a simple and clear-cut taxonomy.

Some of the models described in Part One are, at this stage, of more probabilistic interest than statistical, and it will not escape the reader's attention that the latter is more important to us here. Furthermore, we have in a number of instances refrained from presenting models, or classes of models, in their full generality, in the belief that their nature can more easily be appreciated by examining one or two special cases. In respect of several relevant papers which have come to our attention since this book was drafted, we have in the time available been able to do little more than make a passing reference.

Part Two of the monograph examines in greater detail a single class of models, the hidden Markov models, and their properties. We show how the basic model of that type can be extended and modified to describe not only stationary univariate discrete-valued time series, but also many other types of series including multinomial and multivariate discrete-valued time series. A number of examples of applications of hidden Markov and other selected models are presented in some detail in Chapter 4. The intention is to illustrate the considerable scope and applicability of the available models, especially hidden Markov models. The reader who is seeking a model for a particular discrete-valued series could even begin by browsing through Chapter 4, in order to get a feel for the types of data that can be modelled by hidden Markov models.

Part One of this monograph is the more demanding in terms of the mathematical and statistical background required of the reader. To follow the mathematical details of the models described there, the reader needs to be familiar with the statistical and probabilistic concepts covered in a typical first degree in statistics, including the theory and terminology for (continuous-valued) time series. Less background is required for Part Two, particularly if one wishes only to apply the methods. For that purpose an elementary knowledge of discrete-time Markov chains and of the standard probability distributions such as the binomial, Poisson and multinomial will suffice. While we are aware that simplicity may be in the eye of the beholder, we believe that one of the attractive features of hidden Markov models is their mathematical and conceptual simplicity.

Many statistical questions regarding hidden Markov models remain to be answered. Among these are some of the properties of the parameter estimators used, standard errors and confidence bounds. Goodness-of-fit tests and other diagnostic methods are required. There is a lack of theoretically justified model selection techniques. The (parametric) bootstrap method might eventually provide practical solutions to some of these problems. The properties of the method have yet to be investigated in this context, however. We hope that this monograph will stimulate research towards bridging some of these gaps.

This monograph contains material which has not previously been published, either by ourselves or (to the best of our knowledge) by others. If we have anywhere failed to make appropriate acknowledgement of the work of others, we would be grateful if the reader would draw it to our attention. The applications described in sections 4.4, 4.5 and 4.10 contain *inter alia* some material which first appeared in (respectively) the *South African Statistical Journal*, *Biometrical Journal* and the *International Journal of Epidemiology*. We are grateful to the editors of these journals for allowing us to reuse such material.

Like the list of models, the list of people and institutions to whom we are indebted for encouragement, advice or assistance is long and almost certainly incomplete.

We would particularly like to thank Bernard Silverman for encouraging us to tackle this project. We hope he did not despair of our completing it! Adelchi Azzalini, Peter Guttorp, Peter Lewis, Heinz Linhart, Marty Puterman, Tobias Rydén, Mary Lou Thompson and an anonymous reviewer all read parts or earlier versions of this book and offered comments and advice, all of it valuable. Since we were not always able to follow their advice, the customary disclaimer applies *a fortiori*. David Bowie, Graham Fick, Linda Haines, Len Lerer, Frikkie Potgieter and David Raubenheimer all provided data, and in some cases spent many hours discussing their data with the authors. Shanaaz Hassiem, Günter Kratz and Monika Machoczek provided valuable computing assistance. Ingrid Biedekarken and Ellen Wegener are thanked for their assistance in the preparation of the index.

The first author is grateful to the University of Cape Town Research Fund for financial assistance; to the University of Göttingen for making him very welcome and placing facilities at his disposal during several visits; and to Tony Lawrance and the University of

Birmingham for the opportunity to present and discuss some of the material in this book. The second author wishes to thank the University of Cape Town Research Fund and the Foundation for Research Development (South Africa) for financial assistance.

Finally, we thank the staff of Chapman & Hall for their hard work and patience, and Richard Leigh for his thorough job of copy-editing.

PART ONE

Survey of models

CHAPTER 1

A survey of models for discrete-valued time series

1.1 Introduction: the need for discrete-valued time series models

Many of the statistical time series which occur in practice are by their very nature discrete-valued, although it is often quite adequate, and obviously very convenient, to represent them by means of models based on the normal distribution. Some examples of discrete-valued series are:

  (i) the numbers of defective items found in successive samples taken from a production line;
 (ii) the sequence of wet and dry days at some site;
(iii) the numbers of cases of some notifiable disease in a given area in successive months;
 (iv) the numbers of births, and the numbers of deliveries by various methods, at a hospital in successive months;
  (v) road accident or traffic counts;
 (vi) base sequences in DNA;
(vii) the presence or absence of trading in a particular share on consecutive trading days;
(viii) the numbers of firearm homicides and suicides in successive weeks in a given area; and
 (ix) the behaviour category of an animal observed at regular intervals.

Although in some cases models based on the normal distribution will suffice, this will not always be so. When the observations are categorical in nature, and when the observations are quantitative but fairly small, it is necessary to use models which respect the discrete nature of the data.

Furthermore, there are continuous-valued series in which the observations naturally fall in one of a small number of categories, for instance the series of lengths of eruptions of the Old Faithful geyser analysed by Azzalini and Bowman (1990), and discussed further in section 4.2. In that case most of the observations can be described as 'long' or 'short', with very few eruptions intermediate in length, and the pattern of long and short eruptions is the aspect of most scientific interest. It is therefore natural to treat this series as a binary time series. Another instance of continuous data being treated as discrete is the practice of classifying directional observations into the conventional 16 points of the compass, even if more detailed information is available, because of the familiarity of this classification. The observations of wind direction analysed in section 4.6 are an example of a time series of this kind.

The approach followed here is to seek models consisting of appropriately dependent sequences of discrete random variables, and to develop means of fitting and selecting such models. There is, however, another quite different approach which we will not pursue. This is to recognize that many discrete-valued time series arise as counts in a point process on the line, and to treat them accordingly. For instance, Guttorp (1986) describes the fitting of point process models to binary time series generated by a point process, and Guttorp and Thompson (1990) discuss several methods for estimating point process parameters, in particular the second-order product moment function, from equally spaced observations on the counting process. The approach followed in this book does have the advantage of allowing also for those discrete-valued time series which do not arise as counts of a point process, e.g. categorical series like example (ix) above.

The plan of this book is as follows. The rest of this chapter will survey the models that have been proposed for discrete-valued series, stressing aspects of the models like marginal distributions, correlation structure and (where these have been discussed in the literature) parameter estimation techniques. Although some of these models may seem at this stage to be of more probabilistic interest than statistical, we hope that our survey will be useful to the reader who wishes to know what models are available and, where these have been applied, what they have been applied to. We begin by discussing Markov chains and higher-order Markov chains, in particular the higher-order Markov chain models introduced by Pegram (1980) and generalized by Raftery (1985a). We then summarize (in sections 1.4–1.6) the work on three classes of models that may be described as attempts to provide for discrete-valued time series a broad class of models analogous to the familiar Gaussian ARMA models: models based on mixtures, models based on the idea of thinning a discrete random variable, and certain other bivariate geometric models of (loosely) autoregressive moving average structure. Markov regression models, an extension to the time series context of the ideas of generalized linear models, are discussed next. Section 1.8 presents some relevant examples of parameter-driven processes, i.e. processes in which there is an underlying and unobserved 'parameter process' which determines the distribution of a series of observations. In section 1.9 two contrasting families of state-space models, those of Harvey and of West and Harrison, are described. After surveying further miscellaneous models for discrete-valued series, the chapter ends with a brief account of what has so far been achieved by the various kinds of model.

Chapter 2 begins with a detailed review of certain results on hidden Markov models which are available in the speech-processing literature. Such models, which are examples of parameter-driven processes, have for some time been used in speech-recognition applications, but have only recently been considered as general-purpose models for discrete-valued time series. After this review we introduce the hidden Markov time series models to which the rest of the book is devoted. These models are based on an unobserved stationary Markov chain and either a Poisson or a binomial distribution. Correlation properties are derived, an algorithm for evaluating the likelihood function is described, and direct numerical maximization of the likelihood is proposed as the most practical means of parameter estimation. Marginal, joint and conditional distributions are derived and some applications of these results indicated, e.g. to forecasting and to the treatment of missing data. Reversibility of the observed process is shown to be implied by, but not equivalent to, reversibility of the underlying Markov chain. Finally some remarks are made on the way in which such processes may be used as statistical models.

In Chapter 3 these hidden Markov time series models are extended and modified in various useful ways. In one modification the underlying Markov chain is replaced by a second-order Markov chain, and (*inter alia*) an algorithm for computing the likelihood in this case is derived. In another modification the models based on the binomial distribution are generalized by replacing that distribution by the multinomial. This yields a model for categorical time series. More general multivariate models of several kinds are then discussed, and results on the likelihood function and cross-correlations of such models are derived. Two different methods of incorporating trend or seasonality or dependence on covariates other than time are also introduced in this chapter. One final variation discussed modifies the 'binomial-hidden Markov' model by assuming that the number of trials at each stage is not a known constant but is supplied instead by some further random process.

Chapter 4 presents examples of applications of the models of Chapters 2 and 3 to data of various types and from a variety of disciplines, and makes comparisons with competing models such as Markov chains of order one or higher.

Some remarks on the notation to be used throughout may be helpful at this stage. With few exceptions, the model for the observations will be denoted by $\{S_t\}$. Unless otherwise indicated, vectors are row vectors. Transposition of matrices is denoted by the prime symbol $'$.

1.2 Markov chains

1.2.1 Saturated Markov chains

Since the Markov property is a simple, and mathematically tractable, relaxation of the assumption of independence, it is natural to consider discrete-time Markov chains on a finite state space as possible models for time series taking values in that space. Although some of the models considered later are Markov chains with a specific structure, we confine our attention here to fully parametrized, or saturated, Markov chain models, by which is meant Markov chains which have $m^2 - m$ independently determined transition probabilities when m is the number of states.

We review here the following aspects of stationary Markov chains, on state space $\{1, 2, \ldots, m\}$: the autocorrelation function (ACF), the partial autocorrelation function (PACF), and estimation of the transition probabilities by maximum likelihood. We do so in some detail because of the close links to the 'hidden Markov' models to be introduced in Chapter 2. Unless otherwise noted, Markov chain terminology is taken from Grimmett and Stirzaker (1992). Apart from (possibly) the observations concerning partial autocorrelations, the results of this section (1.2.1) are not new, but some appear not to be readily accessible in the literature. Billingsley (1961) and Chapter 4 of Basawa and Prakasa Rao (1980) present extensive accounts of statistical methods in finite state-space Markov chains.

Let $\{S_t : t \in \mathbf{N}\}$, then, be an irreducible homogeneous Markov chain on the first m positive integers, with transition probability matrix Γ. That is, $\Gamma = (\gamma_{ij})$, where for all states i and j and all times t:
$$\gamma_{ij} = \mathrm{P}(S_t = j \mid S_{t-1} = i).$$
In what follows, unless otherwise indicated, Markov chains are assumed to be homogeneous. By the irreducibility, there exists a unique, strictly positive, stationary distribution, which we shall denote by the vector $\delta = (\delta_1, \delta_2, \ldots, \delta_m)$. Suppose that $\{S_t\}$ is stationary, so that δ is for all t the distribution of S_t.

The irreducibility of $\{S_t\}$ implies also that 1 is a *simple* eigenvalue of Γ and the corresponding right and left eigenvectors are unique up to constant multiples (Seneta, 1981, Theorem 1.5). It then follows that such eigenvectors are multiples of the column vector $\mathbf{1}' = (1, 1, \ldots, 1)'$ and δ respectively.

If we define $v = (1, 2, \ldots, m)$, $V = \mathrm{diag}(v)$ (i.e. the diagonal matrix with v on the principal diagonal), and $\gamma_{ij}(k) = (\Gamma^k)_{ij}$, we have the following results for the mean of S_t and the covariance of S_t and S_{t+k}, for all nonnegative integers k. (It should, however, be noted that such results are not meaningful unless the states $1, 2, \ldots, m$ are quantitative.)

$$\begin{aligned}
\mathrm{E}(S_t) &= \sum_{i=1}^{m} i \delta_i \\
&= \delta v'; \\
\mathrm{E}(S_t S_{t+k}) &= \sum_{i=1}^{m} \sum_{j=1}^{m} ij \delta_i \mathrm{P}(S_{t+k} = j \mid S_t = i) \\
&= \sum_{i,j} (i \delta_i) \gamma_{ij}(k) j \\
&= \delta V \Gamma^k v'; \\
\mathrm{Cov}(S_t, S_{t+k}) &= \delta V \Gamma^k v' - (\delta v')^2.
\end{aligned}$$

Even if Γ is not diagonalizable, some simplification of the expression for the covariance may be achieved by writing Γ in Jordan canonical form, details of which may be found, for instance, in Noble (1969, section 11.6) or in Cox and Miller (1965, pp. 121–122). For our purpose it will be sufficient to note that Γ may be written as $U \Omega U^{-1}$, where U, U^{-1} and Ω are of the following forms:

$U = (\mathbf{1}'\ R)$, $U^{-1} = \begin{pmatrix} \delta \\ W \end{pmatrix}$ and $\Omega = \begin{pmatrix} 1 & \mathbf{0} \\ \mathbf{0}' & \Psi \end{pmatrix}$. (The matrix Ψ is band-diagonal, with the eigenvalues of Γ other than 1 on the diagonal, ones or zeros on the superdiagonal, and zeros elsewhere.) Hence

$$\Gamma^k = U\Omega^k U^{-1} = \mathbf{1}'\delta + R\Psi^k W,$$

and

$$\text{Cov}(S_t, S_{t+k}) = \delta V(\mathbf{1}'\delta + R\Psi^k W)v' - (\delta v')^2 = (\delta V R)\Psi^k(Wv').$$

Since this is true for $k = 0$, we have also the variance of S_t, and hence the ACF of $\{S_t\}$:

$$\rho_k = \frac{\text{Cov}(S_t, S_{t+k})}{\text{Var}\ S_t} = \frac{(\delta V R)\Psi^k(Wv')}{\delta V R W v'}.$$

The resulting expression involves powers up to the kth of the eigenvalues of Γ.

If Γ is diagonalizable, a neater structure emerges. If the eigenvalues (other than 1) of Γ are denoted by $\omega_2, \omega_3, \ldots, \omega_m$, the matrix Ω can be taken to be $\text{diag}(1, \omega_2, \ldots, \omega_m)$, the columns of U are corresponding right eigenvectors of Γ, and the rows of U^{-1} corresponding left eigenvectors. We then have, for $k = 0, 1, 2, \ldots$:

$$\begin{aligned}
\text{Cov}(S_t, S_{t+k}) &= \delta V U \Omega^k U^{-1} v' - (\delta v')^2 \\
&= a\Omega^k b' - a_1 b_1 \\
&= \sum_{i=2}^{m} a_i b_i \omega_i^k,
\end{aligned}$$

where $a = \delta V U$ and $b' = U^{-1} v'$. Hence $\text{Var}(S_t) = \sum_{i=2}^{m} a_i b_i$, and for nonnegative integers k:

$$\rho_k = \text{Corr}(S_t, S_{t+k}) = \sum_{2}^{m} a_i b_i \omega_i^k \Big/ \sum_{2}^{m} a_i b_i.$$

This is a linear combination of the kth powers of the eigenvalues $\omega_2, \ldots, \omega_m$, and (if these eigenvalues are distinct) somewhat similar to the ACF of a Gaussian autoregressive process of order $m - 1$. As will shortly be seen, however, the analogy breaks down when one considers the PACF. We note in passing that $\rho_k = \rho_1^k$ for all nonnegative integers k in the case $m = 2$, and also in certain other cases, e.g. if all the eigenvalues ω_i are equal, or if $a_i b_i = 0$ for all but one value of i. For m equal to 2, ρ_1 is just the eigenvalue of Γ other than 1.

MARKOV CHAINS

Before considering the PACF of a Markov chain, we note first certain general facts concerning partial autocorrelations. For any second-order stationary process $\{S_t\}$, the partial autocorrelation of order k, denoted by ϕ_{kk}, is the correlation of the residuals obtained by regressing S_t and S_{t+k} on the intervening $k-1$ variables. As Lawrance (1976; 1979) demonstrates, this partial correlation is by no means always the same as the conditional correlation of S_t and S_{t+k} given the intervening variables, but equality does hold under normality. (Some authors do not make the distinction clear. For instance Stuart and Ord (1991, section 27.1) describe the conditional correlations but go on to state: 'These are known as the partial correlations.') In order to derive the partial autocorrelations from the autocorrelations ρ_k we may use the standard relation (Brockwell and Davis, 1991, p. 102):

$$\phi_{kk} = |P_k^*|/|P_k|, \tag{1.1}$$

where

$$P_k = \begin{pmatrix} 1 & \rho_1 & \cdots & \rho_{k-1} \\ \rho_1 & 1 & \cdots & \rho_{k-2} \\ \vdots & \vdots & & \vdots \\ \rho_{k-1} & \rho_{k-2} & \cdots & 1 \end{pmatrix}$$

and P_k^* is identical to P_k except that the last column of P_k is replaced by $(\rho_1, \rho_2, \ldots, \rho_k)'$. It should be noted that equation (1.1) holds for any second-order stationary process: see equation (23.4.2) of Cramér (1946), which implies (1.1). In particular, (1.1) does not merely hold for processes with a specific distribution or structure such as Gaussian autoregressive or moving average processes. As Brockwell and Davis remark after their definition 3.4.2, this property provides an alternative definition of the partial autocorrelations.

Now consider the m-state Markov chain $\{S_t\}$ as defined above. By using equation (1.1) it can be verified that, for $m=2$ and any other case in which $\rho_k = \rho_1^k$ for all $k \in \mathbf{N}$, ϕ_{rr} is zero for all r exceeding 1 — as is true also of the standard (Gaussian) AR(1) process. The example below shows, however, that for a three-state Markov chain ϕ_{33} may be nonzero. This is of course quite different behaviour from that of the standard AR(2) process, for which $\phi_{rr} = 0$ for $r \geq 3$.

Example Consider the stationary Markov chain on $\{1, 2, 3\}$ with transition probability matrix as follows:

$$\Gamma = \begin{pmatrix} 0 & 1 & 0 \\ \frac{13}{16} & 0 & \frac{3}{16} \\ 1 & 0 & 0 \end{pmatrix}.$$

The ACF is $\rho_k = \frac{75}{124}(-\frac{3}{4})^k + \frac{49}{124}(-\frac{1}{4})^k$, and the first three autocorrelations are $\rho_1 = -137/248$, $\rho_2 = 181/496$, and $\rho_3 = -1037/3968$. Hence $|P_3^*| = -0.02055$, and ϕ_{33} is nonzero.

In order to estimate the $m^2 - m$ parameters γ_{ij} $(i \neq j)$ of the Markov chain $\{S_t\}$ from a realization $s_1, s_2, \ldots, s_T$, we consider first the likelihood conditioned on the first observation. This is

$$\prod_{i=1}^{m}\prod_{j=1}^{m} \gamma_{ij}^{f_{ij}},$$

where f_{ij} is the number of transitions from state i to state j (and hence $\sum_{i,j} f_{ij} = T - 1$). Since $\gamma_{ii} = 1 - \sum_{k \neq i} \gamma_{ik}$, differentiating the logarithm of this likelihood with respect to γ_{ij} and equating the derivative to zero yields

$$\frac{f_{ii}}{1 - \sum_{k \neq i} \gamma_{ik}} = \frac{f_{ij}}{\gamma_{ij}}.$$

The intuitively plausible estimator $\hat{\gamma}_{ij} = f_{ij}/\sum_{k=1}^{m} f_{ik}$ may thereby be seen to be a conditional maximum likelihood estimator of γ_{ij}. (Note that the assumption of stationarity of the Markov chain was not actually used in the above derivation.) Alternatively, we may use the unconditional likelihood: for a stationary Markov chain $\{S_t\}$ this is the conditional likelihood as above, multiplied by δ_{s_1}, and it or its logarithm may be maximized numerically, subject to nonnegativity and row-sum constraints, in order to estimate the transition probabilities γ_{ij}. In some nontrivial special cases of the two-state Markov chain, explicit expressions are available for the unconditional maximum likelihood estimators: see Bisgaard and Travis (1991).

A saturated Markov chain on m states, having $m^2 - m$ free parameters, cannot be estimated reliably if m is large and the data series is short. Clearly more parsimonious models would be useful.

There are of course applications of Markov chains that assume a specific structure for the transition probability matrix that is appropriate to the observations being modelled. See for instance the Markov chain model of Albert (1994) for a time series of disease severity categories in a disease displaying periods of relapse and

MARKOV CHAINS

periods of remission. In that application a ten-state Markov chain characterized by only eight parameters is used.

1.2.2 A nonhomogeneous Markov chain model for binary time series

The model proposed by Azzalini (1994), which we now describe, is somewhat specialized in that it is designed to deal specifically with binary data. Nevertheless it is an interesting example of how a Markov chain can be used as the basic building-block of a more sophisticated model, in this case one that incorporates covariates in a simple way. Consider a process $S_1, S_2, \ldots, S_T$, assumed to be a nonhomogeneous Markov chain taking on the values 0 and 1, with transition probability matrix, for the transition from S_{t-1} to S_t, given by:

$$\begin{pmatrix} 1 - {}_tp_0 & {}_tp_0 \\ 1 - {}_tp_1 & {}_tp_1 \end{pmatrix}.$$

That is, for $j = 0, 1$:

$$_tp_j = P(S_t = 1 \mid S_{t-1} = j).$$

Given its Markov structure, the distribution of the process can be specified in terms of its (unconditional) means $\theta_t = E(S_t)$ and the odds ratio

$$\psi = \frac{{}_tp_1/(1 - {}_tp_1)}{{}_tp_0/(1 - {}_tp_0)}, \qquad (1.2)$$

which is assumed to be constant over time. (Azzalini indicates, however, that it is possible to relax this assumption and to allow ψ to depend on time-dependent covariates.) On the other hand θ_t is in general permitted to depend on a k-vector x_t of time-dependent covariates as follows:

$$g(\theta_t) = x_t \beta',$$

where $g : (0, 1) \to \mathbf{R}$ is a suitable link function (such as logit or probit), and β is a vector of k parameters. From the relation

$$\theta_t = \theta_{t-1} {}_tp_1 + (1 - \theta_{t-1}) {}_tp_0,$$

and from (1.2), it is possible to obtain the following expression, valid for $t = 2, \ldots, T$, for ${}_tp_j$ in terms of ψ, θ_{t-1} and θ_t:

$$_tp_j = \frac{\delta - 1 + (\psi - 1)(\theta_t - \theta_{t-1})}{2(\psi - 1)(1 - \theta_{t-1})}$$
$$+ j \frac{1 - \delta + (\psi - 1)(\theta_t + \theta_{t-1} - 2\theta_t \theta_{t-1})}{2(\psi - 1)\theta_{t-1}(1 - \theta_{t-1})}, \qquad (1.3)$$

where

$$\delta^2 = 1 + (\psi - 1)\left((\theta_t - \theta_{t-1})^2\psi - (\theta_t + \theta_{t-1})^2 + 2(\theta_t + \theta_{t-1})\right).$$

Formula (1.3) holds for $\psi \neq 1$, and, if $\psi = 1$, ${}_tp_j = \theta_t$. (Note that (1.3) differs slightly from equation (5) of Azzalini (1994), which contains a typographical error: θ_t in the denominator of the last term of that equation should be θ_{t-1}. This error does not affect the rest of his paper, however.) Taking $P(S_1 = 1)$ to be θ_1 completes the specification of the process: $S_1, S_2, \ldots, S_T$ is then a binary (nonhomogeneous) Markov chain with $E(S_t) = \theta_t$ for all t, and with the odds ratio for (S_{t-1}, S_t) in all cases equal to ψ.

For a series of observations $s_1, s_2, \ldots, s_T$, the log-likelihood function for the parameters β and $\lambda = \log \psi$ is:

$$l(\beta, \lambda) = \sum_{t=1}^{T} \left(s_t \mathrm{logit}({}_tp_{s_{t-1}}) + \log(1 - {}_tp_{s_{t-1}}) \right)$$

with ${}_tp_j$ defined by (1.3) for $t > 1$, and ${}_1p_{s_0}$ taken to be θ_1. (Here, as usual, $\mathrm{logit}(x)$ denotes $\log(\frac{x}{1-x})$.) Maximization of this likelihood has to be performed numerically. Although Azzalini gives the first derivatives, he states that it does not seem feasible to obtain expressions for the second-order derivatives. Therefore standard errors of estimates must be obtained by numerical differentiation of the first derivatives. The above likelihood can be modified to cope with missing data (if this is missing at random).

Azzalini describes how models of this kind, which allow the covariates to influence the *marginal* distribution of a given observation, differ from the approach of (for instance) Zeger and Qaqish (1988), in which the covariates influence the conditional distribution given the past. He applies such a model to a study of coronary risk factors in school children in a repeated measures setting, and by means of simulation investigates several theoretical questions not amenable to analytical investigation.

1.3 Higher-order Markov chains

In cases where the observations on a process with finite state space appear not to satisfy the Markov property, one possibility that suggests itself is to fit a higher-order Markov chain, i.e. a model $\{S_t\}$ satisfying the following generalization of the Markov property for some $l \geq 2$:

$$P(S_t \mid S_{t-1}, S_{t-2}, \ldots) = P(S_t \mid S_{t-1}, \ldots, S_{t-l}).$$

An account of such higher-order Markov chains may be found, for instance, in Lloyd (1980, section 19.9). Although such a model is not in the usual sense a Markov chain, i.e. not a 'first-order' Markov chain, we can redefine the model in such a way as to produce an equivalent process which is. If we let $X_t = (S_{t-l+1}, S_{t-l+2}, \ldots, S_t)$, then $\{X_t\}$ is a (first-order) Markov chain on M^l, where M is the state space of $\{S_t\}$. Although some properties are more awkward to establish, no essentially new theory is therefore involved in analysing a higher-order Markov chain rather than a first-order one. For instance, a stationary distribution for $\{S_t\}$, if it exists, may be found by determining the stationary distribution of the Markov chain $\{X_t\}$ and deducing from it the implied marginal distribution for any one of the l components of X_t.

Of course the use of a general higher-order Markov chain (instead of a first-order one) greatly increases the problem of overparametrization: a general Markov chain of order l on m states has $m^l(m-1)$ independent transition probabilities. (Although $\{X_t\}$ is a Markov chain on m^l states, the number of independent transition probabilities is less than $m^{2l} - m^l$ because many of the entries in its transition probability matrix are identically zero.) Pegram (1980) and Raftery (1985a,b) have therefore proposed certain classes of parsimonious models for higher-order chains which are far more suitable for practical application. For $m = 2$ the models of Raftery are equivalent to those of Pegram, but for $m > 2$ those of Raftery are more general. Pegram's models have $m + l - 1$ parameters, and those of Raftery $m(m-1) + l - 1$. This appears to contradict a remark of Li and Kwok (1990), who cite Pegram's work but state that Raftery's proposal 'results in a model that is much more parsimonious than all previously proposed'. If by parsimony one means fewer parameters, clearly Pegram's are the more parsimonious. Raftery's models (for $m > 2$) can, however, represent a wider range of dependence patterns and autocorrelation structures. In both cases an increase of one in the order of the Markov chain requires only one additional parameter.

Raftery's models, which he now terms 'mixture transition distribution' (MTD) models, are defined as follows. The process $\{S_t\}$ takes values in $M = \{1, 2, \ldots, m\}$ and satisfies

$$P(S_t = j_0 \mid S_{t-1} = j_1, \ldots, S_{t-l} = j_l) = \sum_{i=1}^{l} \lambda_i \, q_{j_i j_0}, \qquad (1.4)$$

where $\sum_{i=1}^{l} \lambda_i = 1$, and $Q = (q_{jk})$ is an $m \times m$ matrix with nonnegative entries and row sums equal to one, such that the right-hand side of equation (1.4) is bounded by zero and one for all $j_0, j_1,$ ..., $j_l \in M$. This last requirement, which generates m^{l+1} pairs of constraints nonlinear in the parameters, ensures that the conditional probabilities in equation (1.4) are indeed probabilities, and the condition on the row sums of Q ensures that the sum of these probabilities over j_0 is one. Note that Raftery does not assume that the parameters λ_i are nonnegative. If, however, one does make that assumption, the right-hand side of equation (1.4) is a convex linear combination of probabilities and thereby automatically bounded by zero and one, which renders the nonlinear constraints redundant. The assumption that the parameters λ_i are nonnegative greatly simplifies the computations involved in parameter estimation. If one cannot make that assumption, estimation can be extremely laborious because of the large number of nonlinear constraints on the parameters. However, Raftery and Tavaré (1994) and Schimert (1992) have recently described how one can reduce the number of constraints and thereby make the models much more useful in practice.

In the case $l=1$ this model is a first-order Markov chain with transition probability matrix Q. Another interesting case of the model (1.4) may be obtained by taking

$$Q = \theta I + (1-\theta)\mathbf{1}'\pi,$$

where π is a positive vector with $\sum_{j=1}^{m} \pi_j = 1$. This yields the models of Pegram. The vector π is then the stationary distribution corresponding to Q (although in Pegram's treatment π is actually defined as the limiting distribution, as $t \to \infty$, of S_t).

Raftery proves the following limit theorem, on the assumption that the elements of Q are positive: if π is the stationary distribution corresponding to Q, $S_t = j$ has limiting probability π_j independent of the initial conditions. That is,

$$\lim_{t \to \infty} P(S_t = j \mid S_1 = i_1, \ldots, S_l = i_l) = \pi_j$$

for all $j, i_1, \ldots, i_l \in M$. It is therefore reasonable to restrict one's consideration to stationary models $\{S_t\}$. For such stationary models, Raftery shows that the joint distribution of S_t and S_{t+k} satisfies a system of linear equations similar to the Yule–Walker equations (for which, see e.g. Brockwell and Davis, 1991, p. 239). More precisely, if we define $P(k) = (p_{ij}(k))$ by

$$p_{ij}(k) = P(S_t = i, S_{t+k} = j) \quad i, j, \in M; k \in \mathbf{Z},$$

with $P(0) = \text{diag}(\pi)$ being the case $k=0$, we have for $k \in \mathbf{N}$

$$P(k) = \sum_{g=1}^{l} \lambda_g P(k-g) Q. \tag{1.5}$$

It is not always possible to solve these equations uniquely, but Raftery gives (separately for $l=2$, 3 and ≥ 4) sufficient conditions on the parameters λ_i and q_{jk} for uniqueness. He derives from the equations (1.5) a system of equations for the autocorrelations $\rho_k = \text{Corr}(S_t, S_{t+k})$ which resembles the Yule–Walker equations to some extent and may be solved uniquely in certain special cases only. He considers in detail the autocorrelation behaviour of his model when $m = 3$, $l = 2$, $\pi = \frac{1}{3}\mathbf{1}$ and Q has a special structure implying that the autocorrelations do satisfy precisely a set of Yule–Walker equations, so that

$$\rho_1 = \phi_1 + \phi_2 \rho_1$$
$$\rho_2 = \phi_1 \rho_1 + \phi_2$$

for certain quantities ϕ_i not depending on ρ_1 or ρ_2. It emerges that the set of correlations (ρ_1, ρ_2) allowed by this model, although quite complicated in shape, is contained in and not much smaller than the corresponding set for the usual Gaussian AR(2) models.

Raftery (1985a) fits his models to three data sets, relating to wind power, interpersonal relationships, and occupational mobility. Parameters are estimated by direct numerical maximization of the logarithm of the conditional likelihood, i.e. conditional on the first l observations if l is (as above) the order. Model order selection and comparisons with competing models are done on the basis of the 'Bayesian information criterion' (BIC) of Schwarz (1978). In all three applications the models of Raftery appear to combine parsimony with fidelity to data more successfully than do the alternatives available.

Further applications are presented by Raftery and Tavaré (1994), to DNA sequences, wind direction and patterns in bird-song. In several of the fitted models there are negative estimates of some of the coefficients λ_i.

Adke and Deshmukh (1988) have generalized Raftery's limit theorem to MTD models for higher-order Markov chains on a countable (possibly infinite) state space. In fact they extend the result even further, to similar higher-order Markov models on arbitrary state space. They do, however, make the assumption that the coefficients λ_i are positive.

The merits of various estimation techniques applicable to Raftery's models in several different sampling situations are discussed by Li and Kwok (1990). For a single realization, Raftery's conditional maximum likelihood estimation procedure and a minimum chi-squared procedure are discussed, and compared by means of a simulation experiment involving models in which $m = 3$, $l = 2$ and $\pi = \frac{1}{3}\mathbf{1}$. The two methods produce comparable estimates of Q, but the minimum chi-squared method appears to produce better estimates of λ_1. For 'macro' data, i.e. data which aggregate the observations on several (presumably independent) copies of a Raftery model, a nonlinear least squares method is proposed and investigated by simulation, but the results suggest that the estimators may be inconsistent in some circumstances. Finally Li and Kwok consider unaggregated data from a 'panel' of subjects assumed to follow Raftery models, but with the parameters possibly differing between subjects. On the evidence provided by another simulation experiment Li and Kwok recommend an empirical Bayes technique for the estimation of the parameters for each subject.

Azzalini (1983), in his investigation of maximum likelihood estimation 'of order m', uses a binary second-order Markov chain in order to study the efficiency of such estimation (of order zero or one) in a case where it can be compared with 'exact' maximum likelihood. His conclusion is that the method works well enough to be worth considering in cases where the maximum likelihood estimator is not available. Azzalini and Bowman (1990) report the fitting of a second-order Markov chain model to the binary series they use to represent the lengths of successive eruptions of the Old Faithful geyser. Their analysis, and some alternative models, will be discussed in section 4.2.

A generalization of Raftery's models termed an 'infinite-lag Markov model' is proposed by Mehran (1989). This allows for an infinite number of terms $\lambda_i q_{j_i j_0}$ rather than the l terms appearing in equation (1.4), with $\sum \lambda_i = 1$ as before. In this case, however, the coefficients λ_i are given by some simple decreasing parametric function of i, e.g. $\lambda_i = \varepsilon^{i-1}(1 - \varepsilon)$ for some $\varepsilon \in (0, 1)$. Such models, it is claimed, are particularly useful in circumstances of missing data items or nonconsecutive sampling schemes. A finite sequence is merely treated as an infinite sequence in which all the observations after a certain point are missing. Mehran describes an application to labour statistics collected according to a specific nonconsecutive sampling scheme, the 4-8-4 rotation sampling scheme of the US Current Population Survey.

1.4 The DARMA models of Jacobs and Lewis

The earliest attempt to provide a class of discrete-valued time series models analogous to the familiar Gaussian ARMA models appears to be that of Jacobs and Lewis (1978a–c; 1983). Their two classes of discrete autoregressive moving average models, abbreviated DARMA and NDARMA, are formed by taking probabilistic mixtures of independent and identically distributed (i.i.d.) discrete random variables all having the required marginal distribution. Applications have been published by Buishand (1978), Chang, Kavvas and Delleur (1984a,b), Chang, Delleur and Kavvas (1987), and Delleur, Chang and Kavvas (1989).

In order to specify the models, we need first the following definitions. Let $\{Y_t\}$ be a sequence of i.i.d. random variables on some countable subset E of the real line, with $P(Y_t = i) = \pi(i)$ for all $i \in E$. Let $\{U_t\}$ and $\{V_t\}$ be independent sequences of i.i.d. binary random variables with $P(U_t = 1) = \beta \in [0,1]$ and $P(V_t = 1) = \rho \in [0,1)$. Let $\{D_t\}$ and $\{A_t\}$ be sequences of i.i.d. random variables with

$$P(D_t = n) = \delta_n \quad n = 0, 1, \ldots, N$$
$$P(A_t = n) = \alpha_n \quad n = 1, 2, \ldots, p,$$

where $N \in \mathbf{N}_0$, the set of all nonnegative integers, and $p \in \mathbf{N}$.

The DARMA($p,N+1$) process $\{S_t\}$ is then formed as follows:

$$S_t = U_t Y_{t-D_t} + (1 - U_t) Z_{t-(N+1)} \quad t = 1, 2, \ldots,$$
$$Z_t = V_t Z_{t-A_t} + (1 - V_t) Y_t \quad t = -N, -N+1, \ldots.$$

The process $\{Z_t\}$ is the DAR(p) process, i.e. the discrete autoregressive process of order p, as defined by Jacobs and Lewis (1978c). This process may be described informally as follows: with probability ρ, Z_t is one of the p previous values $Z_{t-1}, \ldots, Z_{t-p}$, and otherwise it is Y_t. The DARMA process $\{S_t\}$ may be described similarly: with probability β, S_t is one of the values $Y_t, \ldots, Y_{t-N}$, and otherwise it equals the 'autoregressive tail' $Z_{t-(N+1)}$. Clearly the above definition of the DARMA process requires specification of the joint distribution of the p-dimensional random vector $(Z_{-N-p}, \ldots, Z_{-N-1})$. In Jacobs and Lewis (1978c) it is shown that there is a stationary distribution for that random vector. If the DAR(p) process $\{Z_t\}$ is started with that stationary distribution, the DARMA process $\{S_t : t \in \mathbf{N}\}$ is then a stationary process with marginal distribution π. We shall henceforth assume that $\{S_t\}$ is indeed stationary.

The cases $\beta = 0$ and $\beta = 1$ of the above general DARMA($p,N+1$)

model are of course simpler in structure than the general model. If $\beta = 0$, we have $S_t = Z_{t-(N+1)}$, i.e. the DAR(p) model already described. If $\beta = 1$, we have $S_t = Y_{t-D_t}$, which is the model termed DMA(N) (not DMA(N+1)) by Jacobs and Lewis (1978a). In this case, S_t is a probabilistic mixture of the $N+1$ i.i.d. random variables $Y_t, Y_{t-1}, \ldots, Y_{t-N}$.

Jacobs and Lewis (1983) derive equations which enable one to find the ACF

$$\rho_k = \mathrm{Corr}(S_t, S_{t+k})$$

of a general DARMA(p, $N+1$) process. The ACF of $\{Z_t\}$, the DAR(p) process, satisfies equations of the same form as the Yule–Walker equations for a Gaussian AR(p) process. The ACF of the DMA(N) process is given by:

$$\rho_k = \begin{cases} \sum_{j=0}^{N-k} \delta_j \delta_{j+k} & k = 1, 2, \ldots, N \\ 0 & k = N+1, \ldots \end{cases}$$

It should be noted that the autocorrelations of any DARMA process are all nonnegative, and do not depend in any way on the marginal distribution π. The marginal distribution itself is completely general, but clearly particular distributions such as the Poisson will be of most interest.

As Jacobs and Lewis (1983) remark, the models of Pegram (1980) are a generalization of finite state-space DAR(p) processes, and unlike the latter they do allow some negative correlation. How strong this negative correlation may be depends on the stationary distribution of the process.

The second class of discrete ARMA models introduced by Jacobs and Lewis (1983) is the class of NDARMA processes ('new' discrete ARMA, presumably). Let $\{Y_t\}$, $\{V_t\}$, $\{D_t\}$ and $\{A_t\}$ be as before. The NDARMA(p,N) process is defined as $\{S'_t\}$, where

$$S'_t = V_t S'_{t-A_t} + (1 - V_t) Y_{t-D_t}.$$

Hence S'_t is, with probability ρ, one of the p previous values S'_{t-1}, $S'_{t-2}, \ldots, S'_{t-p}$. With probability $1 - \rho$, it is one of the $N+1$ quantities $Y_t, Y_{t-1}, \ldots, Y_{t-N}$. As is true also for the DARMA models, special cases yield the DAR and DMA processes. To be specific, the case $\rho = 0$ yields the DMA(N) model, and the case $\delta_0 = \mathrm{P}(D_t = 0) = 1$ yields the DAR(p) model.

Jacobs and Lewis show that NDARMA models have a stationary distribution with marginal distribution π. They derive equations from which the autocorrelations of any stationary NDARMA model may be determined, and note that these too are necessarily nonnegative. As in the case of DARMA models, the correlation structure does not depend on the marginal distribution π, and that marginal distribution is quite general.

Some specific examples of stationary DARMA and NDARMA models are now presented by way of illustration.

DAR(1) As noted above, $\{Z_t\}$ is a DAR(p) process. The DAR(1) process is particularly simple, and satisfies

$$Z_t = V_t Z_{t-1} + (1 - V_t) Y_t.$$

That is,

$$Z_t = \begin{cases} Z_{t-1} & \text{with probability } \rho \\ Y_t & \text{with probability } 1 - \rho. \end{cases}$$

It is therefore a (stationary) Markov chain, with stationary distribution π. The transition probabilities are given by

$$P(Z_t = i \mid Z_{t-1} = k) = \rho \delta_{ki} + (1 - \rho) \pi_i,$$

where δ_{ki} is the Kronecker delta symbol. The ACF is simply ρ^k, for all $k \in \mathbf{N}$.

DMA(1) The DMA(1) model $\{S_t\}$ satisfies

$$S_t = \begin{cases} Y_t & \text{with probability } \delta_0 \\ Y_{t-1} & \text{with probability } \delta_1 = 1 - \delta_0. \end{cases}$$

The ACF is

$$\rho_k = \begin{cases} \delta_0(1 - \delta_0) & k = 1 \\ 0 & k = 2, 3, \dots, \end{cases}$$

and it is shown by Jacobs and Lewis (1978a) that $\{S_t\}$ is reversible. (A random process is said to be reversible if its finite-dimensional distributions are invariant under time-reversal: for some general discussion of reversibility see section 2.9.)

DARMA(1,1) and NDARMA(1,1) The DARMA(1,1) model satisfies

$$S_t = \begin{cases} Y_t & \text{with probability } \beta \\ Z_{t-1} & 1 - \beta, \end{cases}$$

where

$$Z_{t-1} = \begin{cases} Z_{t-2} & \text{with probability } \rho \\ Y_{t-1} & 1 - \rho. \end{cases}$$

The NDARMA(1,1) model $\{S'_t\}$ has

$$S'_t = \begin{cases} S'_{t-1} & \text{with probability} & \rho \\ Y_t & & \delta_0(1-\rho) \\ Y_{t-1} & & (1-\delta_0)(1-\rho). \end{cases}$$

The autocorrelation functions are:

$$\rho_k = \text{Corr}(S_t, S_{t+k}) = \rho^{k-1}(1-\beta)\{\beta(1-\rho) + (1-\beta)\rho\}$$

and

$$\rho'_k = \text{Corr}(S'_t, S'_{t+k}) = \rho^{k-1}\{\rho + (1-\rho)^2 \delta_0(1-\delta_0)\}.$$

Jacobs and Lewis (1983) present graphs of the possible pairs (ρ_1, ρ_2) and (ρ'_1, ρ'_2), from which it is apparent that, although neither set of attainable correlations contains the other, that of the NDARMA-(1,1) process is smaller. By comparison of these graphs with Figure 3.10(b) of Box and Jenkins (1976), one can see that the standard Gaussian ARMA(1,1) model has a much larger set of attainable correlations, mainly because negative correlations are possible.

Jacobs and Lewis (1983) consider the estimation of the autocorrelations of DARMA and NDARMA processes, in particular ρ in the case of a DAR(1) process. On the basis of a simulation study they conclude that the usual sample autocorrelation performs worst among the eight alternatives they present. An *ad hoc* estimator based on properties specific to DARMA and NDARMA models is found to perform well. Many other properties (e.g. the fact that the DARMA(1,N+1) process is ϕ-mixing) and some possible extensions (e.g. to negative correlations and bivariate processes) are described in the papers by Jacobs and Lewis already cited.

We conclude this section instead by describing briefly the applications to hydrological problems that have been reported. Jacobs and Lewis (1983) report the work of Buishand (1978), who used a binary DARMA(1,1) process as a model for sequences of wet and dry days. Chang *et al.* (1984a,b; 1987) and Delleur *et al.* (1989) have applied various DARMA models to sequences of wet and dry days during some 'locally stationary season', to a three-state discretization of daily precipitation amounts, and to estimation of daily runoff. Estimation of the distribution π was done by utilizing the observed runlengths. Chang *et al.* state in these applications that the other parameters were estimated by fitting the theoretical ACF to the sample ACF, by nonlinear least squares: presumably this refers only to the first few terms of the ACF.

1.5 Models based on thinning

A fairly broad class of models is that based on the idea of 'thinning'. Such models are discussed by McKenzie (1985a,b; 1986; 1987; 1988a,b), Al-Osh and Alzaid (1987; 1988; 1991), Alzaid and Al-Osh (1988; 1990; 1993), Du and Li (1991) and Al-Osh and Aly (1992). Dion, Gauthier and Latour (1992) show that these models are simple functionals of a multitype branching process with immigration: this enables them to unify and extend many of the results available for the models based on thinning. In particular they use results on estimation for branching processes to derive corresponding ones for the thinning models.

Although the properties of the models have been studied extensively, it seems that few applications have yet been published. The only application we are aware of is that of Franke and Seligmann (1993), to daily counts of epileptic seizures in one patient.

We introduce the models by considering first the case of those with geometric marginal distribution, and in particular the geometric AR(1) process of McKenzie.

1.5.1 Models with geometric marginal

Let the thinning operation '$*$' (also known as binomial thinning) be defined as follows. If X is any nonnegative integer-valued random variable and $0 \leq \alpha \leq 1$, $\alpha * X$ is defined as $\sum_{i=1}^{X} B_i$, where $\{B_i\}$ is a sequence of i.i.d. binary random variables, independent of X, with $P(B_i = 1) = \alpha$. Conditional on X, therefore, $\alpha * X$ is distributed binomially with parameters X and α. Now suppose that

$$S_t = \alpha * S_{t-1} + R_t, \qquad (1.6)$$

where the innovation R_t is independent of S_{t-1}, $0 \leq \alpha \leq 1$ and S_{t-1} has the geometric distribution with mean $\lambda^{-1} = \theta/(1-\theta)$, i.e. for nonnegative integer k:

$$P(S_{t-1} = k) = (1-\theta)\theta^k. \qquad (1.7)$$

Then S_{t-1} has the 'alternate probability-generating function' (a.p.g.f.)
$$E((1-z)^{S_{t-1}}) = \lambda/(\lambda + z),$$
$\alpha * S_{t-1}$ has a.p.g.f. $\lambda/(\lambda + \alpha z)$, and the condition for S_t to have the same distribution as S_{t-1} is that R_t have a.p.g.f.

$$\frac{\lambda/(\lambda + z)}{\lambda/(\lambda + \alpha z)} = \alpha + (1-\alpha)\frac{\lambda}{\lambda + z}.$$

That is, R_t either is zero (with probability α) or has the geometric distribution (1.7). Equivalently, R_t can be described as the product of (independent) geometric and binary random variables. If R_t satisfies this requirement and S_0 has the distribution (1.7), then S_t will also have that geometric distribution for all nonnegative integers t.

The correlation structure of the stationary sequence $\{S_t\}$ is simple: it can be shown that ρ_k, the autocorrelation of order k, is just α^k, as is the case for the usual continuous-valued AR(1) model

$$S_t = \alpha S_{t-1} + R_t. \tag{1.8}$$

Any random process $\{S_t\}$ satisfying equation (1.6) is a Markov chain, and the transition probabilities for this case are given explicitly in equation (2.8) of McKenzie (1986). The thinning operation '$*$' is a very natural discrete analogue of the scalar multiplication appearing in the model (1.8). Furthermore, the model $S_t = \alpha * S_{t-1} + R_t$ has the possibly useful interpretation that S_t consists of the survivors of those present at time $t-1$ (each with survival probability α) plus R_t new entrants entering between $t-1$ and t.

The above model is the geometric autoregressive process of order one of McKenzie (1985b), described also by Alzaid and Al-Osh (1988). It has been modified and generalized in several directions, mainly by McKenzie and by Al-Osh and Alzaid, and we present a summary of these developments in the rest of section 1.5. These models are designed to have a given marginal distribution (e.g. Poisson, binomial or negative binomial) and dependence structure (autoregressive, moving average or ARMA). It needs to be emphasized, however, that the terms autoregressive and moving average are often used rather loosely in this context, merely to indicate some similarity in form to that of the standard (Gaussian) ARMA models. For example, Alzaid and Al-Osh (1990) state that their integer-valued autoregressive process of order p actually has autocorrelation function similar to that of a standard ARMA$(p, p-1)$ model.

McKenzie (1986) introduces moving average and autoregressive moving average models with geometric marginal, all of which (like the geometric AR(1) model described above) are analogues of the corresponding exponential ARMA models of Lawrance and Lewis (1980), obtained by replacing scalar multiplication by thinning and exponential distributions by geometric. The most general model of this kind described by McKenzie is a geometric ARMA(p, q), but to convey the flavour of these models we present in detail only the geometric ARMA(1,1) process. First let $\{M_t\}$, $\{U_t\}$ and $\{V_t\}$ be independent sequences of i.i.d. random variables, where M_t is geometric with mean λ^{-1}, U_t is binary with $P(U_t=0) = \alpha$, and V_t is binary with $P(V_t=0) = \beta$. Now suppose that W_0 is geometric with mean λ^{-1}, and that

$$S_t = \beta * M_t + V_t W_{t-1}$$

and

$$W_t = \alpha * W_{t-1} + U_t M_t.$$

(Both here and elsewhere it is assumed that each thinning operation is performed independently of all other aspects of the process under consideration, including other thinning operations.) Both W_t and S_t are then geometric with mean λ^{-1}, and $\{S_t\}$ is the geometric ARMA(1,1) process. The kth-order autocorrelation of $\{S_t\}$ is, for positive integers k:

$$\rho_k = \bar{\beta}(\bar{\alpha}\beta + \alpha\bar{\beta})\alpha^{k-1}.$$

(Here $\bar{\beta} = 1 - \beta$, and similarly for other symbols.) Clearly $\{W_t\}$ is a geometric AR(1), and in the special cases $\beta = 0$ and $\alpha = 0$ respectively $\{S_t\}$ is AR(1) and MA(1).

Several other models with geometric marginal have been described. McKenzie (1986) discusses also the geometric analogue of the NEAR(1) process of Lawrance and Lewis (1981), and displays simulations of such a process. It is defined by

$$S_t = (\beta U_t) * S_{t-1} + (1 - V_t + \bar{\alpha}\beta V_t) * M_t,$$

where $\{M_t\}$, $\{U_t\}$ and $\{V_t\}$ are independent sequences of i.i.d. random variables, M_t is geometric, U_t is binary with $P(U_t=1) = \alpha$, and V_t is binary with $P(V_t=1) = \alpha\beta/(1-\bar{\alpha}\beta)$. The case $\alpha=1$ yields the geometric AR(1) process already described above. McKenzie (1985a) gives the geometric analogue of the NEAR(2) model of Lawrance and Lewis (1985).

1.5.2 Models with negative binomial marginal

The geometric distribution is a special case of the negative binomial, and McKenzie (1986) considers also the construction of general negative binomial AR(1) processes. We shall say the random variable S has the negative binomial distribution with shape parameter β and scale parameter λ if for all nonnegative integers k:

$$P(S=k) = \binom{\beta + k - 1}{k} \left(\frac{\lambda}{1+\lambda}\right)^\beta \left(\frac{1}{1+\lambda}\right)^k.$$

The parameters β and λ are assumed only to be positive. Note in particular that the shape parameter β is not necessarily an integer. The a.p.g.f. of such a negative binomial distribution is $(\lambda/(\lambda+z))^\beta$, the Laplace transform of the gamma density with shape and scale parameters β and λ:

$$\lambda^\beta x^{\beta-1} e^{-\lambda x}/\Gamma(\beta) \quad (\text{for } x > 0).$$

This suggests that to define a negative binomial AR(1) process, all one has to do is to replace scalar multiplication by thinning, and gamma distributions by negative binomial, in the gamma AR(1) process of Gaver and Lewis (1980). The model which results is

$$S_t = \alpha * S_{t-1} + R_t,$$

with $0 < \alpha < 1$, S_t having the negative binomial (β, λ) distribution for all t, and the innovation R_t having a.p.g.f.

$$\left(\frac{\lambda + \alpha z}{\lambda + z}\right)^\beta = \left(\alpha + (1-\alpha)\frac{\lambda}{\lambda+z}\right)^\beta.$$

To construct a random variable having this a.p.g.f. for general β (and not merely integer values) requires considerable ingenuity. McKenzie (1987) describes such a construction, based on a shot-noise process, and notes that it is essentially the same as that devised by Lawrance (1982) to solve the corresponding problem for gamma processes.

The complexity of the innovation process led McKenzie, however, to define a different kind of negative binomial process (McKenzie, 1986). This is analogous to the gamma beta AR(1) process described by Lewis (1985), which is a random coefficient autoregression with gamma marginal. The resulting negative binomial AR(1) is defined as follows:

$$S_t = A_t * S_{t-1} + M_t,$$

where A_t has a beta distribution with parameters α and $\beta - \alpha$, M_t has a negative binomial distribution with parameters $\beta - \alpha$ and λ, $0 < \alpha < \beta$, $\lambda > 0$, and A_t, S_{t-1} and M_t are mutually independent. If S_0 has the negative binomial distribution with parameters β and λ, S_t then has that distribution in general, and for all nonnegative

integers k:
$$\rho_k = (\alpha/\beta)^k.$$

Notice that at time t there are three sources of randomness: the (unobserved) random variables A_t and M_t, and the thinning operation.

The special case $\beta = 1$ of this negative binomial AR(1) process provides a geometric AR(1) process other than the one described in section 1.5.1. The innovation of this new geometric AR(1) is neither geometrically distributed nor the product of a geometric and a binary random variable: it is negative binomial with parameters $1 - \alpha$ and λ, where $0 < \alpha < 1$ and $\lambda > 0$.

1.5.3 Models with Poisson marginal

A Poisson AR(1) process is discussed by McKenzie (1985b; 1988b), Al-Osh and Alzaid (1987), and Alzaid and Al-Osh (1988), who treat it as a special case of their INAR(1) model ('integer-valued autoregressive of order 1'). More general Poisson ARMA processes, and a multiple Poisson AR(1), are also introduced by McKenzie (1988b). Al-Osh and Alzaid (1988) discuss various properties of the Poisson case of their INMA(1) and INMA(q) models, INMA standing for 'integer-valued moving average'. Alzaid and Al-Osh (1990) describe properties of the Poisson INAR(p) process, and relate it to the multiple Poisson AR(1) of McKenzie.

McKenzie's Poisson AR(1) process is simply a stationary Poisson solution $\{S_t\}$ of the equation

$$S_t = \alpha * S_{t-1} + R_t,$$

with the usual assumptions that R_t and S_{t-1} are independent and $\{R_t\}$ is a sequence of i.i.d. random variables. What sets the Poisson AR(1) model apart from (e.g.) the geometric AR(1) is that in the Poisson case the marginal and innovation distributions belong to the same family of distributions. More precisely, S_t is Poisson with mean λ if and only if R_t is Poisson with mean $(1 - \alpha)\lambda$: see Al-Osh and Alzaid (1987, section 3). The role of the Poisson distribution relative to the above equation is therefore very much the same as that of the normal distribution relative to the autoregressive equation $S_t = \alpha S_{t-1} + R_t$.

Other interesting properties of the Poisson AR(1) process are that its autocorrelation function is α^k and that it is a reversible

Markov chain with transition probabilities

$$P(S_t=j \mid S_{t-1}=i) = \sum_{k=0}^{j} \{\binom{i}{k}\alpha^k \bar{\alpha}^{i-k}\}\{e^{-\lambda\bar{\alpha}}(\lambda\bar{\alpha})^{j-k}/(j-k)!\}.$$

Here λ is the mean of S_t, and we use the convention that $\binom{i}{k} = 0$ for $k > i$. The regression of S_t on S_{t-1} is linear (and vice versa, by the reversibility), but the variance of S_t given S_{t-1} is not constant with respect to S_{t-1}. This last property is one respect in which the Poisson AR(1) does differ from its Gaussian counterpart. Details may be found in McKenzie (1988b). Al-Osh and Alzaid (1987) describe four techniques for estimating α and the innovation mean in a Poisson AR(1), given a realization $s_0, s_1, \ldots, s_T$. These are Yule–Walker estimation, conditional least squares as proposed by Klimko and Nelson (1978), maximum likelihood conditional on the initial observation, and unconditional maximum likelihood. The first three are compared in a simulation experiment in which the initial observation (s_0) is apparently set equal to the integer part of the process mean. The conclusion drawn is that conditional maximum likelihood performs best of the three, in terms of bias and mean squared error.

The Poisson moving average process of order one, as defined by McKenzie (1988b) and Al-Osh and Alzaid (1988), is a process $\{S_t\}$ satisfying

$$S_t = Y_t + \beta * Y_{t-1}$$

for $0 \leq \beta \leq 1$ and Y_t a sequence of i.i.d. Poisson random variables. If the mean of Y_t is $\lambda/(1+\beta)$, that of S_t is λ. The autocorrelation ρ_k is zero for $k \geq 2$, and $\rho_1 = \beta/(1+\beta)$. What is notable is that the joint distribution of S_{t-1} and S_t is of the same form as in the Poisson AR(1) case discussed above: for both the joint a.p.g.f. is

$$E\left((1-u)^{S_{t-1}}(1-v)^{S_t}\right) = \exp\{-\lambda(u+v-\rho_1 uv)\}.$$

The Poisson moving average process of order q is a natural extension of that of order one:

$$S_t = Y_t + \sum_{i=1}^{q} \beta_i * Y_{t-i},$$

where $0 \leq \beta_i \leq 1$ for all i, $\beta = \sum_{i=0}^{q} \beta_i$, and $\{Y_t\}$ is a sequence of i.i.d. Poisson random variables with mean λ/β. (By convention $\beta_0 = 1$.) The distribution of S_t is Poisson with mean λ, as usual, and the ACF is:

$$\rho_k = \begin{cases} \sum_{i=0}^{q-k} \beta_i \beta_{i+k}/\beta & k = 1, 2, \ldots, q \\ 0 & k > q. \end{cases}$$

The Poisson ARMA$(1, q)$ process is $\{S_t\}$, where

$$S_t = Y_{t-q} + \sum_{k=1}^{q} \beta_k * W_{t+1-k}$$

and

$$Y_t = \alpha * Y_{t-1} + W_t.$$

The sequence $\{W_t\}$ is taken to be an i.i.d. sequence of Poisson random variables with mean $\bar{\alpha}\lambda$, and Y_0 is an independent Poisson of mean λ. It follows that $\{Y_t\}$ is Poisson AR(1), mean λ, and $\{S_t\}$ a stationary sequence of Poisson random variables with mean $(1 + \bar{\alpha}b)\lambda$, where b is defined as $\sum_{k=1}^{q} \beta_k$. If $\alpha = 0$, $\{S_t\}$ is Poisson MA(q); if $\beta_k = 0$ for all k, $\{S_t\}$ is Poisson AR(1). McKenzie (1988b) gives the ACF of the general Poisson ARMA$(1, q)$ process, which has the property that, for $k \geq q$, $\rho_k = \alpha^{k-q}\rho_q$. (Since $\rho_q = (\alpha^q + \bar{\alpha}\sum_{i=1}^{q} \beta_i \alpha^{i-1})/(1 + \bar{\alpha}b)$, there appears to be a minor error in McKenzie's expression for the autocorrelation in the case $k > q$: it seems the first α should be α^q.)

McKenzie also presents results concerning the joint distribution of n consecutive observations from a Poisson ARMA$(1, q)$ process $\{S_t\}$, and uses them to draw conclusions about the reversibility or otherwise of various special cases of the process. The joint distribution of the n consecutive observations is shown to be given by the multivariate Poisson distribution of Teicher (1954). An interesting result is that the directional moments $\text{Cov}(S_t^2, S_{t-k})$ and $\text{Cov}(S_t, S_{t-k}^2)$ are in general equal, although the process may be irreversible. The AR(1), MA(1) and MA(2) processes are in general reversible, but for $q \geq 3$ the MA(q) process may be irreversible.

To define a multiple Poisson AR(1) process McKenzie (1988b) first defines $\alpha * Y$ for Y a random variable taking values in $\mathbf{N}_0$ (the set of all nonnegative integers), and α a vector of probabilities whose sum does not exceed one. This is done by specifying that, conditional on Y, $\alpha * Y$ has a multinomial distribution with parameters Y and α. The operation '*' thus defined is described as multinomial thinning. For a $p \times p$ matrix $A = (\alpha_1 \ \alpha_2 \ \ldots \ \alpha_p)$

and a vector $Y = (Y_1, Y_2, \ldots, Y_p)$ we then define

$$A * Y = \sum_{i=1}^{p} \alpha_i * Y_i,$$

each multinomial thinning here being performed independently. The multiple Poisson AR(1) is defined as a stationary solution $\{X_t\}$ of the equation

$$X_t = A * X_{t-1} + E_t,$$

with $\{E_t\}$ being a sequence of i.i.d. p-dimensional random vectors and each vector X_t consisting of independent Poisson random variables. McKenzie (1988b) presents general results concerning the innovation distribution and correlation structure of such processes, of which the most notable is that the jth component of X_t (as defined above) has ACF $\{\rho_j(k)\}$ satisfying

$$\rho_j(k) = \begin{cases} (A^k)_{jj} & k = 1, 2, \ldots, p \\ \sum_{i=1}^{p} \phi_i \rho_j(k-i) & k \geq p \end{cases}$$

for certain constants $\phi_1, \ldots, \phi_p$. Each component therefore has ARMA$(p, p-1)$ structure, although the orders may obviously be lower for matrices A of a particular structure. This and other aspects McKenzie examines in detail for the case $p = 2$.

Finally, McKenzie indicates possible extensions of the Poisson models in the following directions: compound Poisson marginal distributions, negative correlation (which is precluded by the structure of the above models), and allowance for trend or cyclical behaviour.

Alzaid and Al-Osh (1990) describe the Poisson case of their general INAR(p) process (see our section 1.5.5), and show for $p = 2$ that there is embedded in it a multiple Poisson AR(1) process of the type discussed by McKenzie, i.e. with independence of components. This is the process $\{X_t\}$ defined by the state vector

$$X_t = \begin{pmatrix} S_t \\ \alpha_2 * S_{t-1} \end{pmatrix},$$

where α_2 is the 'coefficient' of S_{t-2} in the defining equation

$$S_t = \alpha_1 * S_{t-1} + \alpha_2 * S_{t-2} + R_t.$$

Because the embedded process is simple in structure, it can then be used to derive properties of $\{S_t\}$ such as the joint distribution of S_{t-1} and S_t and the fact that $\{S_t\}$ is reversible.

1.5.4 Models with binomial marginal

McKenzie (1985b) proposes a binomial AR(1) process $\{S_t\}$ satisfying:
$$S_t = \alpha * S_{t-1} + \beta * (N - S_{t-1}),$$
with S_t binomial with parameters (N, θ) for all t, $0 < \alpha < 1$ and, unless this exceeds one, $\beta = \bar{\alpha}\theta/\bar{\theta}$. (A modification is possible in the case $\bar{\alpha}\theta/\bar{\theta} > 1$.) The ACF is given by $\rho_k = (\alpha - \beta)^k$. The usual construction, involving the equation $S_t = \alpha * S_{t-1} + R_t$, is not possible because, unlike the Poisson, negative binomial (and geometric) distributions, the binomial is not 'discrete self-decomposable' in the sense of Steutel and van Harn (1979).

Al-Osh and Alzaid (1991) construct a rather different class of binomial ARMA models, which we can describe as being based on hypergeometric thinning. To define the models we need the following preliminaries. For a random variable S having a binomial distribution with parameters N and p, let $T(S)$ be a random variable whose distribution conditional on S is given by the hypergeometric distribution with parameters N, s and M:

$$P(T(S){=}k \mid S{=}s) = \frac{\binom{s}{k}\binom{N-s}{M-k}}{\binom{N}{M}},$$

for all appropriate values of k. It then follows that, if S is binomial with parameters (N, p), $T(S)$ is binomial (M, p), and $T(S)$ and $S - T(S)$ are independent.

The binomial AR(1) process of Al-Osh and Alzaid is $\{S_t\}$ defined by
$$S_t = T(S_{t-1}) + R_t,$$
where R_t is independent of S_{t-1} (and of $T(S_{t-1})$) and has a binomial distribution with parameters $N - M$ and p. Hence if S_{t-1} is binomial with parameters (N, p), so also is S_t. Such a binomial AR(1) process, if stationary, has ACF given by $\rho_k = (M/N)^k$ for all $k \in \mathbf{N}_0$, and is in fact a reversible Markov chain. The regression of S_t on S_{t-1} is linear, but the corresponding conditional variance is not constant.

In very similar fashion the same authors define a stationary MA(q) process $\{S_t\}$ satisfying

$$S_t = R_t + \sum_{i=1}^{q} T_i(R_{t-i}),$$

where $\{R_t\}$ is an i.i.d. sequence of random variables binomially distributed with parameters $N-n$ and p, the distribution of $T_i(R_{t-i})$, given R_{t-i}, is hypergeometric with parameters $N-n$, R_{t-i} and N_i, $\sum_{i=1}^{q} N_i = N$, and the various hypergeometric thinnings are performed independently. The ACF is:

$$\rho_k = \begin{cases} \frac{1}{N(N-n)} \sum_{i=0}^{q-k} N_i N_{i+k} & k = 1, 2, \ldots, q \\ 0 & k > q, \end{cases}$$

with N_0 being defined as 1. The process therefore has the usual moving average cut-off property. As is the case with almost all of the models based on thinning, however, the correlations are restricted to being nonnegative.

For the corresponding stationary binomial ARMA(1,q) model $\{S_t\}$ the defining equations are:

$$S_t = Y_{t-q} + \sum_{i=1}^{q} T_i(R_{t+1-i})$$

and
$$Y_t = T(Y_{t-1}) + R_t.$$

This structure is analogous to the construction of McKenzie's Poisson ARMA(1,q) process. The autocorrelation functions of the two processes are very similar too: for $k = 1, 2, \ldots, q$ the explicit expressions for ρ_k are similar in form, and there is in both cases some α such that, for $k \geq q$, $\rho_k = \alpha^{k-q}\rho_q$. The joint distribution of S_{t-1} and S_t is easily derived for the binomial ARMA(1,1) process, and turns out to be of the same symmetric form as for the binomial AR(1) process. Hence the regression of S_t on S_{t-1} (and vice versa) is linear here too. The conclusion of Al-Osh and Alzaid (from the symmetry of the joint distribution of S_t and S_{t-1}) that $\{S_t\}$ is reversible in the binomial ARMA(1,1) case does not seem justified, however.

Al-Osh and Alzaid define also a multiple binomial AR(1) process very similar to the multiple Poisson AR(1) of McKenzie. This model shares with that of McKenzie the property that individual components of the model have ARMA($p, p-1$) correlation structure.

1.5.5 Results not based on any explicit distributional assumption

Some of the results quoted above in the context of particular marginal distributions are far more generally valid. For instance, the property $\rho_k = \alpha^k$ of geometric and Poisson AR(1) models is a property of any INAR(1) process, as defined by Al-Osh and Alzaid (1987): see their equation 3.3. Furthermore, the estimation techniques developed in that paper, although described in detail for the Poisson case only, apply more generally. Further properties of general INAR(1) processes appear in Alzaid and Al-Osh (1988). Results for general INAR(p) processes appear in Alzaid and Al-Osh (1990) and Du and Li (1991), and for general INMA(q) processes in Al-Osh and Alzaid (1988). We now discuss these models and results briefly.

The INAR(p) process $\{S_t\}$ defined by Alzaid and Al-Osh satisfies

$$S_t = \sum_{i=1}^{p} \alpha_i * S_{t-i} + R_t,$$

where $\{R_t\}$ is, as before, a sequence of i.i.d. random variables taking values in $\mathbf{N}_0$, $\sum_{i=1}^{p} \alpha_i < 1$, and the conditional distribution, given S_t, of the random vector

$$(\alpha_1 * S_t, \alpha_2 * S_t, \ldots, \alpha_p * S_t)$$

is multinomial with parameters S_t and $\alpha = (\alpha_1, \ldots, \alpha_p)$, independent of the history of the process: independent, that is, of S_{t-k} and all thinnings thereof, for $k > 0$. This particular structure (which, incidentally, is not the same as the definition given by Du and Li for their INAR(p) process) implies that the correlation structure is similar to that of a standard ARMA($p, p-1$) process, not an AR(p). The definition of Du and Li turns out to imply the standard AR(p) correlation structure. The two papers cited discuss the existence of a stationary or limiting distribution for their respective versions of the INAR(p) process, and derive sufficient conditions for such existence. The key condition in both cases is the requirement that the roots λ of

$$\lambda^p - \alpha_1 \lambda^{p-1} - \cdots - \alpha_{p-1} \lambda - \alpha_p = 0$$

lie inside the unit circle. Al-Osh and Alzaid give a state-space representation of their model which can be used to find its joint distributions. Du and Li discuss Yule–Walker and conditional least squares estimation for their model, and derive a minimum variance prediction formula that is the same as for a standard AR(p).

The general INMA(q) model $\{S_t\}$ of Al-Osh and Alzaid (1988) can be described informally as the Poisson MA(q) of section 1.5.3

minus the Poisson assumption. That is, it satisfies

$$S_t = Y_t + \sum_{i=1}^{q} \beta_i * Y_{t-i},$$

where $\{Y_t\}$ is a sequence of i.i.d. random variables and $0 \leq \beta_i \leq 1$ for all i. The above authors derive, *inter alia*, the ACF of the process, and determine the probability generating function (p.g.f.) of S_t, and the joint p.g.f. of S_t and S_{t+1}, in terms of the p.g.f. of Y_t. The ACF has the cut-off property (after lag q) that one would expect.

1.6 The bivariate geometric models of Block, Langberg and Stoffer

Langberg and Stoffer (1987) and Block, Langberg and Stoffer (1988) present accounts of the properties of certain bivariate models with geometric marginal distributions. These papers do not include any examples of applications of these models to observed time series, and the models appear not to have been pursued in the literature. We therefore consider only the following aspects of such models: marginal distributions, correlation structure, and a brief comparison with other geometric models. In order to ease comparison of this section with the original papers we use the notation of those papers as far as possible, even though it differs from the notation used so far in this chapter.

1.6.1 Moving average models with bivariate geometric distribution

Langberg and Stoffer (1987) introduce a class of bivariate geometric moving average models $\{G(n,m) : n \in \mathbf{Z}\}$, where the positive integer m denotes the order of dependence on the past. Before we define the models, however, it is worth noting that the above authors use the term 'geometric distribution' to mean a distribution on the *positive* integers with probability mass function of the form $p(1-p)^{k-1}$ ($k \in \mathbf{N}$) for some $p \in (0,1]$. The mean is then p^{-1} and the p.g.f. $ps/\{1-(1-p)s\}$. (Others, e.g. McKenzie, use the term for a distribution on the nonnegative integers.) A bivariate geometric distribution is any bivariate distribution with geometric marginals.

Let the column vectors $M(n) = (M_1(n), M_2(n))'$ be i.i.d. bivariate geometric random vectors, with common mean $(p_1^{-1}, p_2^{-1})'$. Let the column vectors $N(n) = (N_1(n), N_2(n))'$ be independent bivariate geometric with mean vectors $(\alpha_1(n)/p_1, \alpha_2(n)/p_2)'$, independent of all $M(n)$. (We suppose that $p_i \leq \alpha_i(n) \leq 1$ for $i = 1, 2$ and all n.) Let $(J_1(n,j), J_2(n,j))$ be independent random vectors, independent of all $M(n)$ and $N(n)$, such that $J_i(n,j)$ is a binary random variable with probability $1 - \alpha_i(n - j + 1)$ of equalling 1. (This differs from the probability appearing in Langberg and Stoffer (1987): they have $1 - \alpha_i(n)$. As is discussed further below, their definition appears to be in error.) Define

$$U_q(n,j) = \begin{pmatrix} \prod_{k=q}^{j} J_1(n,k) & 0 \\ 0 & \prod_{k=q}^{j} J_2(n,k) \end{pmatrix},$$

that is, $U_q(n,j) = J(n,q)J(n,q+1)\cdots J(n,j)$, where:

$$J(n,k) = \begin{pmatrix} J_1(n,k) & 0 \\ 0 & J_2(n,k) \end{pmatrix}.$$

For simplicity $U_1(n,j)$ is written $U(n,j)$, and equals the product $J(n,1)J(n,2)\cdots J(n,j)$. The bivariate geometric moving average model of order m, sometimes abbreviated to BGMA(m), is now defined by

$$G(n,m) = \sum_{r=0}^{m} U(n,r)N(n-r) + U(n,m+1)M(n-m).$$

It is worth noting here that there is no 'cross-coupling': the first component of $G(n,m)$, for instance, depends only on the first component of the vectors $N(n-r)$ and $M(n-m)$, and not on the second. In order to show that $G(n,m)$ has the required distribution, viz. a bivariate geometric distribution with mean vector $(p_1^{-1}, p_2^{-1})'$, Langberg and Stoffer consider the more general random vector $H_q(n,m)$, defined as

$$\sum_{r=0}^{m} U_q(n, r+q-1)N(n-r-q+1) + U_q(n, m+q)M(n-m-q+1).$$

The vector $G(n,m)$ is the special case $q = 1$. They state that $H_q(n,m)$ has for all q the required distribution, but both steps of their inductive proof of this result (Lemma 3.7) seem to require the definition given above for $J_i(n,j)$ rather than the one they give.

The case $m = 1$, i.e. the moving average model of order one, will suffice to illustrate the structure of these processes:

$$G(n, 1) = N(n) + J(n, 1)N(n - 1) + J(n, 1)J(n, 2)M(n - 2)$$
$$= N(n) + J(n, 1)\{N(n - 1) + J(n, 2)M(n - 2)\}.$$

Notice that, with the definition given above, all three of the following random vectors have a bivariate geometric distribution with mean vector $(p_1^{-1}, p_2^{-1})'$: $M(n-2)$, the vector in curly brackets, and $G(n, 1)$.

Langberg and Stoffer give, in their equation (3.17), a general expression for the covariance structure at lag h, i.e. for the 2×2 matrix

$$\Gamma^m(n, h) = (\text{Cov}(G_i(n, m), G_j(n + h, m))).$$

It should be noted that this is not in general a symmetric matrix. The case $m = 1$ of their result yields *inter alia* the following properties of the moving average of order one, with $\Xi_N(n)$ denoting the covariance matrix of the vector $N(n)$:

$$\Gamma^1(n, 1) = \Xi_N(n) \begin{pmatrix} 1 - \alpha_1(n+1) & 0 \\ 0 & 1 - \alpha_2(n+1) \end{pmatrix};$$

and, for $h = 2, 3, \ldots$:

$$\Gamma^1(n, h) = \mathbf{0}.$$

A similar cut-off property holds for higher-order models, that is $\Gamma^m(n, h) = \mathbf{0}$ for $h > m$.

Langberg and Stoffer also define a similar moving average model of infinite order and discuss its properties.

1.6.2 *Autoregressive and autoregressive moving average models with bivariate geometric distribution*

Block, Langberg and Stoffer (1988) define a bivariate geometric autoregressive model of general order m, denoted by BGAR(m), and two bivariate geometric autoregressive moving average models of orders m_1 and m_2, both denoted by BGARMA(m_1, m_2). In their definition of the BGAR process, the column vectors $M(n)$ and $N(n)$ are as in section 1.6.1. The binary random variables $J_i(n, m)$ are, however, defined differently from before. We suppose now that the $2m$-component random vectors $J(n)$, defined by

$$J(n) = (J_1(n, 1), \ldots, J_1(n, m), J_2(n, 1), \ldots, J_2(n, m)),$$

satisfy the requirements

$$P((J_i(n,1),\ldots,J_i(n,m))=\mathbf{0}) = \alpha_i(n)$$

and

$$\sum_{j=1}^{m} P((J_i(n,1),\ldots,J_i(n,m))=e_j) = 1 - \alpha_i(n)$$

for $i = 1, 2$ and all n, where e_j has its jth component equal to one and the other $m - 1$ components equal to zero. That is, $(J_i(n,1),\ldots,J_i(n,m))$ is either all zeros (with probability $\alpha_i(n)$), or a single one and $m - 1$ zeros (with probability $1 - \alpha_i(n)$). We suppose that the vectors $J(n)$ are independent of each other and of all vectors $M(n)$ and $N(n)$. Define also, for $q = 1, 2, \ldots, m$:

$$C(n,q) = \begin{pmatrix} J_1(n,q) & 0 \\ 0 & J_2(n,q) \end{pmatrix}.$$

The BGAR(m) process is then defined by

$$G(n) = \begin{cases} M(n) & n = 0, 1, \ldots, m-1 \\ \sum_{q=1}^{m} C(n,q) G(n-q) + N(n) & n = m, m+1, \ldots \end{cases}.$$

It follows, by induction on n, that $G(n)$ has a bivariate geometric distribution with mean vector $(p_1^{-1}, p_2^{-2})'$.

If we assume that $\alpha_i(n) = \alpha_i$ for all n and $i = 1$ and 2, so that the marginal processes $\{G_i(n)\}$ are stationary, we can derive equations for the autocorrelations

$$\rho_i(k) = \mathrm{Corr}(G_i(n), G_i(n+k)) \quad (i=1,2;\ n=m,m+1,\ldots;\ k \in \mathbf{N}_0).$$

If we define $\gamma_i(q) = P(J_i(n,q) = 1)$ for $i = 1, 2$ and $q = 1, \ldots, m$, so that $\sum_{q=1}^{m} \gamma_i(q) = 1 - \alpha_i$, the result is the following equation of Yule–Walker form, for $k = m, m+1, \ldots$:

$$\rho_i(k) = \gamma_i(1)\rho_i(k-1) + \gamma_i(2)\rho_i(k-2) + \cdots + \gamma_i(m)\rho_i(k-m).$$

Assuming $\alpha_i(n) = \alpha_i$ for all n, while sufficient for marginal stationarity, is not sufficient for stationarity of the bivariate process. Block et al. describe also a bivariate-stationary BGAR(1) model, and derive its covariance structure, i.e. the auto- and cross-covariances at a given lag $k \in \mathbf{N}_0$. The covariance structure turns out to be very similar to (but slightly simpler than) the covariance structure of a bivariate Gaussian AR(1) model: compare equation (4.12) of Block et al. with equation (9.4.7) of Priestley (1981). The model under discussion is simpler because there is no cross-coupling in its definition.

The two bivariate geometric ARMA models $\{L(n)\}$ defined by Block *et al.* may be described briefly as follows. Let $\{G(n)\}$ be a BGMA(m_1) model with mean vector $(\beta_1/\delta_1, \beta_2/\delta_2)'$, and $\{H(n)\}$ a BGAR(m_2) with mean vector $(\delta_1^{-1}, \delta_2^{-1})'$. Let $U_1(n)$ and $U_2(n)$ be binary variables with $P(U_i(n)=1) = 1-\beta_i$. Then with appropriate independence assumptions we may define

$$L_i(n) = G_i(n) + U_i(n)H_i(n)$$

for $i = 1$ and 2, and so obtain a process $\{L(n)\}$ such that $L(n)$ has a bivariate geometric distribution. Alternatively, we may exchange the mean vectors of $G(n)$ and $H(n)$, and define instead

$$L_i(n) = H_i(n) + U_i(n)G_i(n).$$

Various other properties of all the bivariate geometric models defined above are considered in detail in the two papers cited, especially positive dependence properties. In the context of the present work, however, it may be more interesting to consider an example of one of the models and compare it with similar time series models with geometric marginal.

Let us therefore consider the first component, $\{G_1(n)\}$, of a BGAR(1) model $\{G(n)\}$. The fundamental property of the sequence $\{G_1(n)\}$ is that for $n \in \mathbf{N}$

$$G_1(n) = J_1(n,1)G_1(n-1) + N_1(n),$$

where the three random variables on the right-hand side are independent, $J_1(n,1)$ is binary, and $N_1(n)$ is a geometric 'innovation', but with mean differing from that of $G_1(n-1)$ and $G_1(n)$ (except in a certain trivial case). The most important difference between this model and the geometric AR(1) of McKenzie is that $G_1(n-1)$ is not here 'thinned': either the whole of $G_1(n-1)$ is included in $G_1(n)$ (along with $N_1(n)$) or it is not included at all. In McKenzie's model, each of the 'individuals' present at time $n-1$ is considered *separately* and, with the appropriate survival probability, included among the survivors to time n.

The geometric DAR(1) process of Jacobs and Lewis (see section 1.4) may be described as a process, with geometric marginal, such that the value at time n is either the same as the value at time $n-1$ (with probability ρ) or else an innovation having exactly the geometric distribution required as marginal. Therefore the construction of such a process also differs from that of $\{G_1(n)\}$. All three of these 'geometric AR(1)' models do, however, have ACFs of geometrically decreasing form α^k — as does yet another geometric

AR(1), the model with negative binomial innovation described at the end of section 1.5.2.

1.7 Markov regression models

A class of models that are both versatile and convenient to apply has been described by Zeger and Qaqish (1988). These models, known as Markov regression models, are essentially an extension to the time series context of the ideas of generalized linear models and quasi-likelihood. Their utility is certainly not confined to discrete-valued observations, although here we consider such applications only. The principal advantages of this class of models are their flexibility, especially as regards the inclusion of covariates, and the ease with which parameter estimates may be computed by standard software.

To illustrate the flexibility, we begin with an example of a binomial model incorporating time trend, seasonal variation and autoregression. Suppose that S_t, the observation at time t, is the number of 'successes' in n_t trials and that, conditional on the history $S^{(t-1)} = \{S_k : 1 \leq k \leq t-1\}$, S_t has a binomial distribution with parameters n_t and p_t, where for some positive integers q and r:

$$\begin{aligned}
\text{logit } p_t &= \log\left(\frac{p_t}{1-p_t}\right) \\
&= \alpha_1 + \alpha_2 t + \gamma_1 \sin(2\pi t/r) + \gamma_2 \cos(2\pi t/r) \\
&\quad + \beta_1(S_{t-1}/n_{t-1}) + \cdots + \beta_q(S_{t-q}/n_{t-q}).
\end{aligned} \quad (1.9)$$

This model is similar to, but slightly more general than, the 'linear logistic autoregression' of Cox (1981), which is based on a Bernoulli rather than a binomial distribution and lacks the terms representing trend and seasonality. Model (1.9) makes allowance for dependence of the success probability on the proportion of successes at each of the previous q time points, time trend and r-period seasonality. Clearly it is straightforward to add further terms to the above expression for logit p_t to allow for the effect of any further covariates thought relevant. (There is a problem if, for example, $n_{t-1} = 0$, and it is not clear how one should modify the model to allow for such a case.) The model is an example of what Cox (1981) terms 'observation-driven' models: it is observation-driven in the sense that the distribution of the observation at a given time is specified in terms of the observations at earlier times. In the next section we shall consider some examples of a rather different class of models, the processes described by Cox as 'parameter-driven'.

Suppose we have available a realization $\{s_t : 1 \leq t \leq T\}$ of model (1.9). The likelihood of $s_{q+1}, \ldots, s_T$, conditional on the first q observations, is just

$$\prod_{t=q+1}^{T} \binom{n_t}{s_t} p_t^{s_t}(1-p_t)^{n_t-s_t}.$$

To estimate the parameters $\alpha_1, \alpha_2, \gamma_1, \gamma_2, \beta_1, \ldots, \beta_q$ we may maximize the conditional likelihood with respect to these parameters by performing a logistic regression of S_t on t, $\sin(2\pi t/r)$, $\cos(2\pi t/r)$, and $S_{t-1}/n_{t-1}, \ldots, S_{t-q}/n_{t-q}$. Programs such as GLIM, GENSTAT, S-PLUS and SAS may conveniently be used for this purpose.

More generally, let $\{S_t\}$ be the observed time series, and X_t a (row) vector of p covariates. Let D_t, the 'information set', consist of the past observations $S_1, \ldots, S_{t-1}$ and present and past covariate vectors $X_1, \ldots, X_t$. Let the conditional mean of S_t, given D_t, be denoted by μ_t and suppose that the corresponding conditional variance is $\phi V(\mu_t)$. (This relation between mean and variance is the usual quasi-likelihood assumption except that we are dealing here with conditional rather than marginal means and variances.) Suppose further that some link function g of the conditional mean is linear in the current covariates X_t and in q known functions of the past observations and covariates. That is, for some functions f_i, often of S_{t-i} and X_{t-i} only, and some vectors β and $\theta = (\theta_1, \ldots, \theta_q)$ of parameters, the following is true:

$$g(\mu_t) = \beta X_t' + \sum_{i=1}^{q} \theta_i f_i(D_t). \qquad (1.10)$$

A simple example of such a model, given by Zeger and Qaqish (1988), is that for binary outcomes

$$g(\mu_t) = \text{logit } \mu_t = \beta X_t' + \sum_{i=1}^{q} \theta_i s_{t-i}. \qquad (1.11)$$

In this case $V(\mu) = \mu(1-\mu)$ and $\phi = 1$.

For models satisfying (1.10) the estimation of the vector $\gamma = (\beta, \theta)$ from a realization $s_1, \ldots, s_T$ may be accomplished by solving the (conditional) quasi-likelihood estimating equations

$$\sum_{t=q+1}^{T} \frac{\partial \mu_t}{\partial \gamma}(s_t - \mu_t)/V(\mu_t) = \mathbf{0} \qquad (1.12)$$

by iterative weighted least squares, e.g. by using GLIM. These equations result from equating to zero the 'quasi-score function', i.e. the derivative with respect to γ of the log of the quasi-likelihood function. If we define

$$Z_t = (X_t, f_1(D_t), \ldots, f_q(D_t))$$

and note that

$$\frac{\partial \mu_t}{\partial \gamma} = \frac{1}{\dot{g}(\mu_t)} Z_t,$$

where $\dot{g}$ denotes the derivative of g, we see that the estimating equations (1.12) are equivalent to

$$\sum_{t=q+1}^{T} \frac{1}{\dot{g}(\mu_t)} Z_t (s_t - \mu_t)/V(\mu_t) = \mathbf{0}.$$

When $\dot{g}(\mu) = 1/V(\mu)$, g is described as a 'canonical' link, and the equations reduce to

$$\sum_{t=q+1}^{T} Z_t (s_t - \mu_t) = \mathbf{0}.$$

Since the term 'canonical link' is usually defined in the context of a distribution assumed to belong to the exponential family, the above usage is a slight extension of the usual. The link function $g(\mu) = \text{logit } \mu$ in the model (1.11) is an example of a canonical link.

If $\hat{\gamma}$ denotes the parameter estimates thus obtained, the distribution of $\sqrt{T}(\hat{\gamma} - \gamma)$ converges (under appropriate regularity conditions) to multivariate normal with mean zero and covariance matrix ϕW^{-1}, where

$$W = \lim_{T \to \infty} T^{-1} \sum_{t=q+1}^{T} Z_t' \dot{g}(\mu_t)^{-2} V(\mu_t)^{-1} Z_t.$$

If g is a canonical link, W reduces to

$$\lim_{T \to \infty} T^{-1} \sum_{t=q+1}^{T} Z_t' V(\mu_t) Z_t.$$

The scale parameter ϕ may in general be estimated by

$$\hat{\phi} = T^{-1} \sum_{t=1}^{T} \hat{a}_t^2,$$

where $\hat{a}_t$ is the residual $(s_t - \hat{\mu}_t)/V(\hat{\mu}_t)^{1/2}$.

The above treatment is essentially that of Zeger and Qaqish (1988) and Li (1991), except that it ignores two complications discussed by the former authors. Firstly, the functions f_i may in fact not be known completely, but may depend on the parameters β. Zeger and Qaqish describe several useful models of this kind, and generalize the estimation algorithm to allow for this possibility. Secondly, it is necessary in some cases to estimate parameters other than those included in β and θ. A modification of the estimation algorithm is also possible in that case. Zeger and Qaqish describe in detail an application of their methods, but to a continuous-valued series, which is not directly of interest here.

Li (1991) introduces two methods of assessing the adequacy of a Markov regression model. One is based on the residual autocorrelations

$$\hat{C}_k = T^{-1} \sum_{t=k+1}^{T} \hat{a}_t \hat{a}_{t-k}/\hat{\phi},$$

which are shown to have asymptotically a multivariate normal distribution with zero mean. The other method, which is simpler to use, is based on the score function (more pedantically, the quasi-score). The score statistic derived has a chi-squared asymptotic distribution. Li describes a simulation study of the behaviour of the score statistic in respect of three discrete-valued models without covariates, and reports 'quite reasonable' results. An extension of the models of Zeger and Qaqish to allow for moving average terms as well as autoregressive is described by Li (1994).

Fahrmeir and Kaufmann (1987) and Kaufmann (1987) describe models for categorical time series which are similar to the models of Zeger and Qaqish except in two respects. Firstly, they are not set in a quasi-likelihood context. Secondly, in order to represent the q categories of possible outcome, the response is a vector of $q-1$ binary variables, and not a scalar as in the case of the models of Zeger and Qaqish. Kaufmann proves very general results relating to the asymptotic normality, consistency and efficiency of maximum likelihood estimators for these models (*inter alia*), results which are in fact the basis of some of the conclusions drawn by Zeger and Qaqish. Fahrmeir and Kaufmann discuss the testing of linear hypotheses on the parameters of their models by means of three different (but asymptotically equivalent) statistics, and indicate some tests of particular practical significance, e.g. a test of independence of parallel series and a test of nonsignificance of the covariates. They report also a simulation study of the finite-sample properties of the estimators, in particular in the case of a binary-response model with first-order Markov dependence and two binary covariates. The results are promising, and they conclude that the asymptotic distributions seem to be sufficiently accurate for inference purposes, even at moderate sample sizes.

Liang and Zeger (1989) and Zeger and Liang (1991) discuss multivariate Markov regression models in which at least some of the components of the vector of responses may be discrete. The first paper relates specifically to multivariate binary series, and to conditional logistic regression models for such series which specify the distribution of one response variable given the past values of that response, current and past values of the other responses, and current values of covariates. The conditional log-likelihood approach is computationally burdensome, and a 'pseudo-likelihood' is maximized instead. An application to health-care utilization by 300 families enrolled in a health-maintenance plan in Maryland is described.

The second paper cited above refers to situations (and models) in which the vector of responses may include both discrete and continuous components. For each response some link function of the conditional mean is taken to be linear in the current covariates, the current values of the other responses, and the past values of all the responses. The conditional variance of the jth response is assumed to be proportional to some known function V_j of the conditional mean. The approach already described, based on the quasi-likelihood 'estimating equations', may be generalized to provide an estimate of the vector of parameters. The distribution of this estimate converges, as before, to multivariate normal with the correct mean. An application to infectious diseases and vitamin A deficiency in Indonesian pre-school children is described in detail by Zeger and Liang.

Albert et al. (1994) present a nonlinear extension of the models and methods of Zeger and Qaqish, and apply it in a study of the use of magnetic resonance imaging, repeated at regular intervals, as a measure of disease activity in multiple sclerosis.

The 'linear contagion model' of Holden (1987) for aircraft hijackings in the USA in the years 1968–1972 can also be described as a Markov regression model. Conditional on the history, the number of hijacking events in period t is taken to be a Poisson random variable with mean λ_t, where λ_t is a linear function of the numbers of events in preceding periods, possibly with similar linear contributions from the histories of several covariates. The weight structure chosen by Holden implies that any past hijacking event makes a positive contribution to λ_t, a contribution that is initially low, with time increases to a peak, and then dies out. A similar model for inhibition rather than contagion would have negative weights rather than positive.

In section 4.4 we present an application of logistic-linear autoregressive models, similar to (1.9), to data relating to births in successive months at Edendale hospital in Natal, South Africa. In that section the number of deliveries by Caesarean section is modelled as a proportion of the total number of deliveries, and related, *inter alia*, to the corresponding proportion in the previous month.

1.8 Parameter-driven models

We now consider models of the kind described by Cox (1981) as 'parameter-driven'. Conditional on some unobserved 'parameter process', the observations in such a model are independent, with distribution determined by the current state of the parameter process. For instance, the conditional distribution of an observation could be Poisson, with mean λ_1 or λ_2 depending on whether a two-state parameter process is in state 1 or state 2. Alternatively, the mean could be provided by some positive-valued process having (say) gamma or lognormal marginal distribution. As Cox explains, there may be circumstances in which a parameter-driven model is appropriate to, or even strongly suggested by, the data. Consider a very long binary series consisting almost entirely of zeros, but with occasional bursts of ones fairly close together. Such observations suggest some underlying process which occasionally changes to a state in which one or zero is possible, and then reverts to its normal state, in which only zeros are possible.

In a parameter-driven process (as defined above), the only source of dependence and autocorrelation between observations is the parameter process. If the parameter process happens to consist of independent random variables, then the observations are independent, with distribution compounded by the marginal distribution of the parameter process.

The models of Keenan (1982) for binary time series are examples of parameter-driven processes: they can be described briefly as follows. Let $\{X_t\}$ be an unobserved (completely) stationary process with state space $\mathbf{R}$, and let the 'response function' $F : \mathbf{R} \to [0, 1]$ be monotone. Suppose that, conditional on $\{X_t\}$, the random variables $\{S_t\}$ are independent, with (conditional) distribution specified by

$$P(S_t=1) = 1 - P(S_t=0) = F(X_t).$$

One model $\{S_t\}$ considered in particular is that which results if $\{X_t\}$ is a Gaussian process with mean zero and F the distribution function of a normal distribution with mean zero and variance σ^2. In that case explicit expressions are found for the distributions of two and three consecutive observations in the process $\{S_t\}$. The joint distribution of more than three consecutive observations is not available in closed form. The one-step-ahead forecast distribution based on observations $s_1, s_2, \ldots, s_T$ must therefore be approximated if $T > 2$. Keenan presents and compares six different approximations to this distribution under the assumption that $\{X_t\}$ is a Gaussian AR(1). The use of the sample autocorrelations of $\{S_t\}$ to estimate the parameters of $\{X_t\}$ and the parameter σ in the response function is also explored in some generality.

The models of Kedem (1980) for 'clipped' series, although similar to those of Keenan, do not really meet the definition of parameter-driven, because the original series, as well as the clipped version, is assumed to be available. Some details will be presented in section 1.10.

Azzalini (1982) discusses a model in which the parameter process $\{\theta_t\}$ is the gamma AR(1) process of Gaver and Lewis (1980) and provides the mean for observations which are conditionally Poisson. The three parameters of the gamma AR(1) are initially assumed known. A fairly simple recursive filtering procedure is proposed, i.e. a procedure for estimating the unobserved state variable θ_t from $S^{(t)}$, the history up to and including time t. This Azzalini compares with a second filtering algorithm, which is very similar to the Kalman filter: a simulation experiment suggests that the first method is slightly better in terms of squared-error loss. Another possibility investigated by Azzalini is to use the first 50 observations (say) of a realization to estimate the three parameters by the method of moments, and thereafter to use the observations for both parameter estimation and filtering. Again the first filtering method seems slightly the better of the two.

Zeger (1988) introduces a class of parameter-driven loglinear regression models for time series of counts. They are defined as follows. Suppose that, conditional on the unobserved process $\{\varepsilon_t\}$, the observed process $\{S_t\}$ is a sequence of independent counts, with S_t having conditional mean and variance both equal to $\exp(\beta X_t')\varepsilon_t$.

The (row) vector X_t contains the values at time t of the p covariates, and the coefficients β are the quantities of interest. Suppose further that $\{\varepsilon_t\}$ is a stationary process with $E(\varepsilon_t)=1$ and

$$\text{Cov}(\varepsilon_t, \varepsilon_{t+\tau}) = \sigma^2 \rho_\varepsilon(\tau)$$

for all $\tau \in \mathbf{N}_0$. The marginal properties of $\{S_t\}$ are then:

$$\mu_t = E(S_t) = \exp(\beta X_t')$$
$$\text{Var}(S_t) = \mu_t + \sigma^2 \mu_t^2;$$

and, for $\tau \in \mathbf{N}$ but not $\tau = 0$,

$$\text{Corr}(S_t, S_{t+\tau}) = \frac{\rho_\varepsilon(\tau)}{[1+(\sigma^2\mu_t)^{-1}]^{1/2}[1+(\sigma^2\mu_{t+\tau})^{-1}]^{1/2}}.$$

This model is similar to the conventional Poisson-loglinear model in that $\log \mu_t = \beta X_t'$, but different in that $\{S_t\}$ displays autocorrelation and overdispersion relative to the Poisson:

$$\text{Var}(S_t) = \mu_t(1+\sigma^2\mu_t) > \mu_t.$$

Furthermore, the extent of this overdispersion depends on μ_t. For estimation of β, from observations $s = (s_1, \ldots, s_T)'$, Zeger proposes that the solution $\hat{\beta}$ of the following quasi-likelihood estimating equations be obtained by iterative weighted least squares:

$$D'V^{-1}(s-\mu) = \mathbf{0},$$

where

$\mu = (\mu_1, \ldots, \mu_t)'$,
$V =$ the covariance matrix of $S_1, \ldots, S_T$, and
$D = \left(\frac{\partial \mu_i}{\partial \beta_j}\right)$.

Unlike the usual case of independence, or even the case of the Markov regression models, V is here nondiagonal. Inversion of V is difficult, and Zeger suggests as an alternative to $\hat{\beta}$ an estimate $\hat{\beta}_R$ based on V_R, an approximation to V which is easier to invert. Both algorithms require estimates of σ^2 and other nuisance parameters: Zeger proposes moments estimators for these. Asymptotic results for $\hat{\beta}$ and $\hat{\beta}_R$ are stated, and the efficiency of $\hat{\beta}_R$ relative to $\hat{\beta}$ is discussed for some simple models. A model of the above type is fitted to US polio incidence data for the years 1970–1983, and used to estimate the time trend in incidence. Comparisons are made with two conventional loglinear models which assume independence.

An application of Zeger's models, to the relationship between sudden infant death syndrome (SIDS) and environmental temperature, is described by Campbell (1994). He analyses three data sets, and finds a strong negative relationship between temperature and deaths atttributable to SIDS. Such deaths appear to increase 2–5 days after a decrease in temperature.

The hidden Markov models of Chapters 2 and 3 are parameter-driven processes, and fairly straightforward to use as statistical models. In such models the parameter process is a Markov chain, or possibly a higher-order Markov chain.

1.9 State-space models

The 'state-space' time series models of firstly Harvey and secondly West and Harrison do not fit neatly into the framework of either of the two broad categories described above, observation-driven and parameter-driven processes. Both of these state-space approaches are applicable to discrete-valued series, and have been developed at some length in the books of Harvey (1989) and West and Harrison (1989), as well as in the papers of Harvey and Fernandes (1989a,b) and West, Harrison and Migon (1985), among others. We therefore give here only brief accounts of these approaches. A comparative review of the above books has been provided by Fildes (1991).

We illustrate the 'structural' models of Harvey by means of an example, the model in which the observations are Poisson with mean given by a gamma process. Similar models are available for other distributions: see Harvey (1989, section 6.6). Suppose then that the observation S_t has a Poisson distribution with mean μ_t, and that the process $\{\mu_t\}$ evolves as follows. Conditional on the information available at time $t-1$, i.e. the history $S^{(t-1)}$, μ_{t-1} has a gamma distribution with shape and scale parameters a_{t-1} and b_{t-1} respectively. (These parameters are as defined in section 1.5.2.) Given the same information, μ_t is taken to have a gamma distribution with parameters

$$a_{t|t-1} = \omega a_{t-1}$$
$$b_{t|t-1} = \omega b_{t-1}$$

for some $\omega \in (0,1]$. Updating this prior information with the observation S_t, we obtain the posterior for μ_t, which is gamma with parameters given by

$$a_t = a_{t|t-1} + S_t$$
$$b_t = b_{t|t-1} + 1.$$

(This makes use of the properties of the gamma distribution as natural conjugate prior for the Poisson.) Using the result of Lewis, McKenzie and Hugus (1989) on the beta-gamma transformation, we can see that the transition from μ_{t-1} (given $S^{(t-1)}$) to μ_t (also given $S^{(t-1)}$) could equivalently be described by

$$\mu_t = \omega^{-1}\mu_{t-1}\eta_t$$

for some η_t independent of μ_{t-1} and having the beta distribution with parameters ωa_{t-1} and $(1-\omega)a_{t-1}$. Two consequences of the definition of $\{\mu_t\}$ are that:

$$\mathrm{E}\left(\mu_t \mid S^{(t-1)}\right) = a_{t|t-1}/b_{t|t-1} = \mathrm{E}\left(\mu_{t-1} \mid S^{(t-1)}\right)$$

and

$$\mathrm{Var}\left(\mu_t \mid S^{(t-1)}\right) = a_{t|t-1}/b_{t|t-1}^2 = \omega^{-1}\mathrm{Var}\left(\mu_{t-1} \mid S^{(t-1)}\right).$$

The process $\{\mu_t\}$ is initialized with $a_0 = b_0 = 0$, although that yields an improper distribution until the first time (τ) at which there is a nonzero observation. The likelihood of observations $S_{\tau+1}$, ..., S_T, conditional on $S^{(\tau)}$, is obtained as a product of negative binomial probabilities, and may be maximized in order to estimate the 'hyperparameter' ω.

The one-step-ahead forecast function yielded by this approach can be shown to be a weighted mean of observations in which the weights decline exponentially, and given approximately by an exponentially weighted moving average if the sample is large. Harvey shows also that the k-step-ahead forecast function, for $k > 1$, is identical to that for one step ahead.

The dynamic generalized linear model of West and Harrison is an extension both of their own (normal-theory) dynamic linear model and of the usual 'static' generalized linear model. It is an extension of the first in that nonnormal distributions (including discrete) are provided for, and of the second in that time-varying parameters are incorporated in the model. It is a more fully Bayesian approach than that of Harvey, and attempts to solve a more general problem. In such a model the scalar observation S_t has an exponential family distribution. The associated canonical parameter has the natural conjugate prior distribution and is linked to the state vector θ_t (a column vector) by

$$g(\eta_t) = F_t\theta_t,$$

where the monotone function g and the (row) vector F_t are assumed known. The function g is often taken to be the identity.

MISCELLANEOUS MODELS   47

The state vector evolves according to

$$\theta_t = G_t \theta_{t-1} + \omega_t,$$

where the matrix G_t is known and ω_t has mean zero and known covariance matrix W_t. A useful summary of the recursions involved in obtaining the posterior for η_t, updating the state vector, and obtaining the k-step-ahead forecast distribution for $\{S_t\}$, is presented by West, Harrison and Migon (1985, section 3.3).

Their applications include the forecasting of a series of monthly sales data which exhibit both trend and seasonal variation; a series of half-hourly counts of telephone calls at a university telephone exchange; and a series of bounded counts used to study the relationship between television advertising and consumer awareness, and to measure the effectiveness of advertising campaigns.

Two more recent contributions to the literature on the use of state-space models and methods in the analysis of discrete-valued time series are those of Kashiwagi and Yanagimoto (1992) and Singh and Roberts (1992).

1.10 Miscellaneous models

An autoregressive model for binary (i.e. zero–one) series that seems to have been used more as a building-block for other models than as a time series model in its own right is that of Kanter (1975). The (stationary) model of order N satisfies:

$$S_t = \begin{cases} S_{t-k} \oplus U_t & \text{with probability } p_k, k = 1, 2, \ldots, N \\ U_t & \text{with probability } p, \end{cases}$$

where $\{U_t\}$ is a sequence of i.i.d. binary random variables, $\oplus$ denotes addition modulo 2, and $p + \sum_{k=1}^{N} p_k = 1$. McKenzie (1981) uses Kanter's model, and a similar binary moving average model, to generalize several EARMA processes (for positive continuous-valued series) and the DMA(1) process of Jacobs and Lewis, by replacing the independent binary mixing in the definition of such processes by dependent. He thereby extends the range of correlations possible in those processes. In the course of so doing he indicates how to generalize Kanter's model to one of general ARMA structure (rather than merely autoregressive).

Another binary sequence sometimes mentioned as a possible time series model is the model of Klotz (1973) for Bernoulli trials with dependence. This is, however, just a stationary two-state Markov chain parametrized by $p = \delta_2$ and $\lambda = 1 - \gamma_2$, where δ_2 and γ_2 are as in section 1.2.1. Klotz suggests certain easily computed *ad hoc* estimators of p and λ, and applies them to rainfall data. Devore (1976) claims that Klotz's *ad hoc* procedure is unnecessary because estimation based on either the full unconditional likelihood or the likelihood conditional on the first observation is very straightforward. He states in particular that the unconditional maximum likelihood estimates can be found by solving a quadratic equation. However, as pointed out by Bisgaard and Travis (1991), and the authors cited by them, the estimates are in fact given by the solution of a cubic equation.

The book of Kedem (1980) is mainly concerned with binary series derived from some observed continuous-valued series $\{Z_t\}$ by 'clipping' or 'hard limiting', i.e. by a transformation

$$S_t = \begin{cases} 0 & Z_t < a \\ 1 & Z_t \geq a. \end{cases}$$

Typically the observations Z_t are taken to be generated by a stationary Gaussian autoregressive process, and the object is to estimate parameters by using only the clipped series $\{S_t\}$. Such a procedure may well be attractive in circumstances of very fast data acquisition. Defining F to be the step function

$$F(x) = \begin{cases} 0 & x < a \\ 1 & x \geq a, \end{cases}$$

we see that, given $\{Z_t\}$, $P(S_t = 1) = F(Z_t)$. Apart from the distinction already mentioned in section 1.8, the clipped series is therefore a model of the same general type as those investigated by Keenan. As it is a rather special-purpose model we shall not consider it further here.

Blight (1989) has derived properties of certain discrete-valued series formed by superposing a number of independent renewal processes which all start with an 'event' at time zero and have inter-event intervals taking positive integer values only. One such renewal process generates a binary series in obvious fashion: at each time point $t \in \mathbf{N}$ there is either one event or none. If one superposes N independent processes of this kind, not necessarily identical in nature, the result is a series $\{S_t\}$ taking values in $\{0, 1, \ldots, N\}$. If the N processes are independent copies of the same renewal process, the distribution of S_t is binomial. In this case (of identical processes being superposed) Blight derives an ARMA representation for the process $Z_t = S_t - \mathrm{E}(S_t)$ on the assumption that the probability generating function of the inter-event intervals is rational or polynomial. For instance, if that generating function is

$$\left(\frac{pz}{1-(1-p)z}\right)^2,$$

then $\{Z_t\}$ satisfies the ARMA representation

$$Z_t - (1-2p)Z_{t-1} = a_t - \beta a_{t-1},$$

where $\beta = (1-p)^{-1}\{(1-p+p^2) - p\sqrt{2-2p+p^2}\}$ and $\{a_t\}$ is an uncorrelated 'noise' sequence. Some generalizations are indicated, e.g. to the case of the superposed processes being nonidentical, and to the case of the p.g.f. being neither rational nor polynomial.

Under the heading of 'Walsh–Fourier analysis' Stoffer (1985; 1987; 1990; 1991) has provided an extensive account of what has been termed a sequency domain approach to the analysis of time series, especially discrete-valued and categorical series. Although the emphasis in this book falls on time domain analysis, we present here a brief description of this alternative approach. The fundamental idea of Walsh–Fourier analysis may be stated as follows. There are series of observations, e.g. discrete-valued, in which the signal can be modelled better by the superposition of square waveforms than sinusoidal. The Walsh functions, like the trigonometric functions employed in Fourier analysis, form a complete orthogonal sequence on [0,1). They are similar in oscillation and many other properties to the trigonometric functions, but unlike the trigonometric functions, which vary smoothly, they assume two values only: +1 and −1. Observations displaying sharp discontinuities and a limited number of levels can therefore be represented better by the Walsh functions than by trigonometric. Stoffer discusses the 'Walsh–Fourier' analogues of various standard techniques of Fourier analysis, and in Stoffer (1987) shows how Walsh–Fourier analysis can be used, for instance, to estimate transition probabilities in a 'macro model' which aggregates several independent copies of a Markov chain and superimposes a (discrete-valued) noise term.

1.11 Discussion

It has been stated as recently as 1988 that 'time series models for a sequence of dependent discrete random variables ... are rare' (Al-Osh and Alzaid, 1988). The reader of this chapter might be forgiven, however, for concluding that published applications of many of the models that do exist are rarer. For instance, few applications of the 'marginally specific' models of sections 1.4–1.6 seem to have appeared yet. There are even those who question the usefulness of an approach which is based on marginal distributions rather than on conditional distributions given the history of the process: see the comments of Diggle and Westcott (1985), although it should be noted that those comments were made in the context of positive-valued series, not discrete-valued. What Diggle and Westcott were suggesting is the use of models which are specified in observation-driven form, rather than models designed to have a given marginal distribution and dependence structure, some of which are rather contrived. But given the success of the Gaussian ARMA models in the analysis of continuous-valued series, it is surely sensible to find out whether any similar approach can be made to work for at least some kinds of discrete-valued series. Difficulties in obtaining the likelihood do seem to be an obstacle to the use of many of the marginally specific models.

The Markov regression models are better developed and easier to apply: in some cases existing software can be used directly. They are applicable to a variety of discrete-valued time series. The state-space models of section 1.9 present a more or less Bayesian approach to the modelling problem under discussion: more in the case of the dynamic generalized linear model of West and Harrison, less in the case of Harvey's structural time series models. While Harvey believes that there is nothing in his proposed methods to which a classical statistician could object (Harvey and Fernandes, 1989a), there are those who are uneasy about the mix of classical and Bayesian notions involved (Winkler, 1989). The apparent complexity of the structure of the dynamic generalized linear model does present a barrier to its easy application. Furthermore, the methods of West and Harrison are, in common with many other Bayesian proposals, open to the criticism that a large number of quantities are assumed known *a priori*. Nevertheless the great flexibility of the dynamic generalized linear model makes it potentially a very useful tool, and its application has certainly been taken further than is true of many of the models surveyed in this chapter.

Parameter-driven models, as pointed out by Cox and Snell (1989, p. 101), can be difficult to use as a basis for the analysis of data. Compare, for instance, the parameter-driven models of Zeger (1988) with the observation-driven models of Zeger and Qaqish (1988). Since observation-driven models are not always appropriate, it is worthwhile to try to develop a class of parameter-driven models which are parsimonious, flexible and fairly easy to apply. It will be argued in Part Two that hidden Markov models are such a class of models. The ease with which they can be modified or extended in order to accommodate many different kinds of data is a major advantage. Among the types of time series data for which hidden Markov models can be used are: unbounded counts (i.e. Poisson-like observations), bounded counts (binomial- or multinomial-like observations), multivariate discrete observations, categorical observations, vector observations with some components discrete and some continuous, and discrete observations displaying trend or seasonality or dependence on covariates other than time.

PART TWO

Hidden Markov models

CHAPTER 2

The basic models

2.1 Introduction

Suppose a series of unbounded counts is observed, e.g. counts of epileptic seizures in one patient on successive days, or counts of homicides in a given area in successive weeks. The simplest model that can be considered is that the counts are generated by a homogeneous Poisson process. This implies that the counts are independent identically distributed Poisson random variables and therefore have the property that the variance is equal to the mean. However, many such time series of counts observed in practice are overdispersed relative to the Poisson distribution (i.e. the variance exceeds the mean), and also exhibit serial dependence.

An alternative model, which allows for overdispersion, is as follows. Suppose that each count is generated by one of two Poisson distributions, with means λ_1 and λ_2, where the choice of mean is made by another random mechanism, which we call the parameter process. For example, λ_1 is selected with probability δ_1, and λ_2 with probability $\delta_2 = 1 - \delta_1$. In this case it can be shown that the variance of the counts is equal to the mean, $\delta_1\lambda_1 + \delta_2\lambda_2$, plus a nonnegative term, $(\lambda_1 - \lambda_2)^2 \delta_1 \delta_2$.

If the parameter process is a series of independent random variables, the counts are also independent. But if the parameter process is taken to be a Markov chain, the resulting process of counts allows for serial dependence in addition to overdispersion. It is a simple example of the class of models discussed in this part of the book, namely hidden Markov models.

Such probabilistic functions of a Markov chain have for some time been used in engineering applications, for example as the acoustic model in speech processing: see Levinson, Rabiner and Sondhi (1983) or Juang and Rabiner (1991). There is a substantial engineering literature on the methodology and applications of hidden Markov models, often more complex models than the ones we discuss in this book: see for instance Elliot, Aggoun and Moore (1995) and journals such as the *IEEE Transactions on Signal Processing* and *IEEE Transactions on Speech and Audio Processing*. The name 'hidden Markov model' is apparently due to L.P. Neuwirth (Poritz, 1988).

Similar models have also been used in genetics and biochemistry: see Thompson (1983), Churchill (1989; 1992), Guttorp, Newton and Abkowitz (1990), Baldi *et al.* (1994) and Krogh *et al.* (1994). Juang (1985) states in passing that hidden Markov models 'have been found to be extremely useful for stock market behavior', and a similar claim is made by Kemeny, Snell and Knapp (1976, p. 468). Zucchini and Guttorp (1991) apply hidden Markov models to the modelling of the wet–dry sequence at one or several sites — that is, they use them as multivariate binary time series models. Albert (1991) and Le, Leroux and Puterman (1992) use them to model time series of epileptic seizure counts, and Leroux and Puterman (1992) apply them to the pattern of movements of a foetal lamb.

Further recent applications are the analysis of patch clamp recordings of the current in an ion channel in a cell membrane (Fredkin and Rice, 1992a,b; Ball and Rice, 1992, p. 195) and the modelling of disease activity in multiple sclerosis (Albert *et al.,* 1994).

In a hidden Markov model an underlying and unobserved sequence of states follows a Markov chain with finite state space, and the probability distribution of the observation at any time is determined only by the current state of that Markov chain. The main object of this chapter is to develop such models as general-purpose models for discrete-valued time series. First, however, we review relevant aspects of the theory of hidden Markov models as applied in speech processing. We describe in detail both the 'Baum–Welch re-estimation algorithm', the algorithm widely used to solve the estimation (or 'training') problem in such applications, and the Viterbi algorithm, a dynamic programming method for determining the most likely sequence of states in the underlying Markov chain. It will not disturb the continuity greatly if the reader should omit section 2.2, as the models and techniques we propose do not require any of the results of that section for their derivation or implementation. Nevertheless it is interesting to compare the various approaches to estimation in hidden Markov models, and this will be done where appropriate.

The models we propose, which differ slightly from those used in speech processing and from those of Albert and those of Leroux and Puterman, are defined in section 2.3. Thereafter their correlation properties are derived, and the key property of these models which makes them feasible as practical statistical models is discussed. This is the property that the likelihood of even a very long sequence of observations can be computed sufficiently fast to enable parameters to be estimated by direct numerical maximization of that likelihood. In section 2.7 this estimation technique is compared in detail with that of Leroux and Puterman, which (like the Baum–Welch algorithm) is an implementation of the EM algorithm. The distributional properties of the models proposed are derived in section 2.6, and the statistical implications of these results are indicated, in particular their relevance to forecasting and the treatment of missing data. Section 2.9 discusses the reversibility or otherwise of the models, and shows that reversibility of the observed process is not equivalent to that of the underlying Markov chain. In section 2.10 some concluding remarks are made, in particular on the way in which the models may be used in practice.

2.2 Some theoretical aspects of hidden Markov models in speech processing

One use of hidden Markov models in isolated word recognition (as opposed to continuous speech recognition) may be described briefly as follows. We wish to be able to recognize an utterance known to come from a known vocabulary of V words. (The term 'word' is to be interpreted broadly, as meaning the language unit being modelled, not necessarily a word in the usual sense. It could, for instance, be some subword unit.) Each utterance gives rise to an observation sequence (the acoustic signal) $s_1, s_2, \ldots, s_T$, which is regarded as a realization of length T of some random process $\{S_t : t \in \mathbf{N}\}$ of finite state space. The process $\{S_t\}$ is taken to be generated by two probabilistic mechanisms: firstly, an unobserved (homogeneous) Markov chain $\{C_t\}$ on m states representing the configurations of the vocal tract at successive instants of time; and secondly, a set of probability distributions, one for each state of $\{C_t\}$, that produce the observations from a finite set of n possibilities. Such a hidden Markov model can therefore be characterized by the distribution of C_1 (denoted by δ), the transition probability matrix of the Markov chain (Γ), and the $n \times m$ matrix Π of probabilities defined by

$$\pi_{si} = P(S_t = s \mid C_t = i).$$

The 'training problem', which must be solved separately for each of the V words in the vocabulary, is that of finding satisfactory estimates of δ, Γ and Π, given an observation sequence (or sequences) known to come from an utterance of a particular word. The 'classification problem' is that of deciding which word in the vocabulary a given observation sequence corresponds to. One way of performing the classification is to compute the probability (under the model derived at the training stage) of each word, given the observation sequence, and then to choose that word in the vocabulary which maximizes this probability.

If the parameters δ, Γ and Π are to be estimated by maximum likelihood and the classification performed by the above method, the classification and training problems involve the evaluation, and maximization with respect to δ, Γ and Π, of the likelihood

$$P(S_1 = s_1, S_2 = s_2, \ldots, S_T = s_T),$$

which will be denoted by L_T. In this context the likelihood is usually evaluated by the 'forward–backward' algorithm, and maximum likelihood estimates of the parameters computed by the 'Baum–Welch re-estimation algorithm'. (The terminology varies, however: some authors use the term 'forward–backward algorithm' for the estimation algorithm.) The Baum–Welch algorithm was developed by L.E. Baum and his co-workers in a series of papers published between 1966 and 1972: Baum and Petrie (1966), Baum and Eagon (1967), Baum and Sell (1968), Baum et al. (1970), and Baum (1972). The name of Welch seems to appear only as joint author (with Baum) of a paper listed by Baum et al. (1970) as submitted for publication. The algorithm is in fact an early example of an algorithm of EM type. We shall first derive the forward–backward and Baum–Welch algorithms, and then indicate how the latter algorithm fits into the EM framework.

We therefore consider $\{S_t\}$ and $\{C_t\}$ as described above, and assume explicitly for all positive integers T that, given $C^{(T)} = \{C_t : t = 1, 2, \ldots, T\}$, the random variables $S_1, \ldots, S_T$ are mutually independent. The distribution of S_t given $C^{(T)}$ is assumed to depend only on C_t and to be given by

$$P(S_t = s \mid C_t = i) = \pi_{si}.$$

The independence assumption is not usually mentioned in the speech-processing literature, but it (or an equivalent assumption) is implicit in the derivation of the forward–backward and Baum–Welch algorithms. We shall give a rather complete account of these derivations, because some of the results needed seem not to have been proved in the published literature. Baum et al. (1970), for instance, refer to the apparently unpublished paper by Baum and Welch for certain of the results. Furthermore, the expository article of Juang and Rabiner (1991) gives only a brief account (on p. 256) of the relation between the Baum–Welch and EM algorithms, and it therefore seems useful to discuss that relation more fully here.

We begin by stating four properties of the model $\{S_t\}$ that will be needed. We shall often use a somewhat abbreviated notation, in which for instance the event that $S_t = s_t$ is denoted by S_t.

Firstly, for $t = 1, 2, \ldots, T$:
$$P(S_1, \ldots, S_T \mid C_t) = P(S_1, \ldots, S_t \mid C_t) P(S_{t+1}, \ldots, S_T \mid C_t). \quad (2.1)$$

(In the case $t = T$ we use the convention that $P(S_{t+1}, \ldots, S_T \mid C_t) = 1$.) Secondly, for $t = 1, 2, \ldots, T-1$:
$$P(S_1, \ldots, S_T \mid C_t, C_{t+1}) = P(S_1, \ldots, S_t \mid C_t) P(S_{t+1}, \ldots, S_T \mid C_{t+1}). \quad (2.2)$$

Thirdly, for $1 \leq t \leq l \leq T$:
$$P(S_l, \ldots, S_T \mid C_t, \ldots, C_l) = P(S_l, \ldots, S_T \mid C_l). \quad (2.3)$$

Finally, for $t = 1, 2, \ldots, T$:
$$P(S_t, \ldots, S_T \mid C_t) = P(S_t \mid C_t) P(S_{t+1}, \ldots, S_T \mid C_t). \quad (2.4)$$

Given the structure of the models, none of these properties is surprising, and we defer the proofs to Appendix A.

The forward-backward algorithm consists essentially of the computation of the 'forward probabilities' $\alpha_t(i)$ and 'backward probabilities' $\beta_t(i)$, so called because they require respectively a forward and a backward pass through the data. These probabilities are used in the Baum–Welch algorithm, and are defined as follows for all states i of the Markov chain, and all t from 1 to T:
$$\alpha_t(i) = P(S_1 = s_1, \ldots, S_t = s_t, C_t = i)$$
and
$$\beta_t(i) = P(S_{t+1} = s_{t+1}, \ldots, S_T = s_T \mid C_t = i).$$
(The convention noted above implies that $\beta_T(i) = 1$ for all i.) From these definitions and property (2.1) we have, for $t = 1, 2, \ldots, T$:
$$\alpha_t(i) \beta_t(i) = P(C_t = i) P(S_1, \ldots, S_t \mid C_t = i) P(S_{t+1}, \ldots, S_T \mid C_t = i)$$

$$= \mathrm{P}(C_t{=}i)\,\mathrm{P}(S_1,\ldots,S_T \mid C_t{=}i)$$
$$= \mathrm{P}(S_1,\ldots,S_T, C_t{=}i), \qquad (2.5)$$

and

$$\sum_{i=1}^{m} \alpha_t(i)\beta_t(i) = \mathrm{P}(S_1,\ldots,S_T) = L_T. \qquad (2.6)$$

Hence, if we can evaluate the forward and backward probabilities for all t, we have available T different ways of computing the likelihood. For instance, setting $t = T$ yields $L_T = \sum_{i=1}^{m} \alpha_T(i)$, the formula usually quoted in the speech-processing literature.

In order to find all $\alpha_t(i)$ and $\beta_t(i)$ we note that $\beta_T(i) = 1$ and

$$\alpha_1(i) = \mathrm{P}(C_1{=}i)\,\mathrm{P}(S_1{=}s_1 \mid C_1{=}i) = \delta_i\,\pi_{s_1 i},$$

and we use these values to start the two recursions derived below, which are valid for $1 \leq t \leq T - 1$. Firstly, by using property (2.2), we have

$$\alpha_{t+1}(j)$$
$$= \sum_{i=1}^{m} \mathrm{P}(S_1,\ldots,S_{t+1}, C_t{=}i, C_{t+1}{=}j)$$
$$= \sum \mathrm{P}(C_t{=}i, C_{t+1}{=}j)\,\mathrm{P}(S_1,\ldots,S_{t+1} \mid C_t{=}i, C_{t+1}{=}j)$$
$$= \sum \mathrm{P}(C_t{=}i)\,\gamma_{ij}\,\mathrm{P}(S_1,\ldots,S_t \mid C_t{=}i)\,\mathrm{P}(S_{t+1} \mid C_{t+1}{=}j)$$
$$= \sum \mathrm{P}(S_1,\ldots,S_t, C_t{=}i)\,\gamma_{ij}\,\pi_{s_{t+1} j}$$
$$= \left(\sum_{i=1}^{m} \alpha_t(i)\,\gamma_{ij}\right) \pi_{s_{t+1} j}.$$

Secondly, by using property (2.3) (with $l = t + 1$) and property (2.4), we have

$$\beta_t(i)$$
$$= \sum_{j=1}^{m} \mathrm{P}(S_{t+1},\ldots,S_T, C_t{=}i, C_{t+1}{=}j)/\mathrm{P}(C_t{=}i)$$
$$= \sum \mathrm{P}(S_{t+1},\ldots,S_T \mid C_t{=}i, C_{t+1}{=}j)$$
$$\qquad\qquad \times \mathrm{P}(C_t{=}i, C_{t+1}{=}j)/\mathrm{P}(C_t{=}i),$$
$$= \sum \mathrm{P}(S_{t+1},\ldots,S_T \mid C_{t+1}{=}j)\,\gamma_{ij}$$
$$= \sum \mathrm{P}(S_{t+1} \mid C_{t+1}{=}j)\,\mathrm{P}(S_{t+2},\ldots,S_T \mid C_{t+1}{=}j)\,\gamma_{ij}$$

$$= \sum_{j=1}^{m} \pi_{s_{t+1}j}\beta_{t+1}(j)\,\gamma_{ij}.$$

As Levinson et al. (1983) point out, the above results can be stated more succinctly in matrix notation. If we define, for all t from 1 to T, the vectors

$$\alpha_t = (\alpha_t(1), \alpha_t(2), \ldots, \alpha_t(m))$$

and similarly

$$\beta_t = (\beta_t(1), \beta_t(2), \ldots, \beta_t(m)),$$

the result (2.6) for the likelihood can be written as

$$L_T = \alpha_t \beta_t' \text{ for all } t.$$

The recursions for the forward and backward probabilities are given by

$$\alpha_{t+1} = \alpha_t B_{t+1}$$

and

$$\beta_t' = B_{t+1} \beta_{t+1}'$$

if we define $B_t = \Gamma \lambda(s_t)$ and $\lambda(s)$ is the $m \times m$ diagonal matrix with ith diagonal element equal to π_{si}. These recursions start from $\alpha_1 = \delta\lambda(s_1)$ and $\beta_T = \mathbf{1}$. The following explicit expressions for the forward and backward probabilities are therefore available:

$$\alpha_t = \delta\lambda(s_1) B_2 B_3 \cdots B_t$$

and

$$\beta_t' = B_{t+1} B_{t+2} \cdots B_T \mathbf{1}'.$$

With the convention that an empty product of matrices is the identity matrix, these expressions hold for all t from 1 to T inclusive. A matrix expression for the likelihood will be derived independently in section 2.5, and its uses described in sections 2.6 and 2.7.

We now discuss the rationale of the three 're-estimation formulas' which constitute the Baum–Welch algorithm for improving estimates of δ, Π and Γ. We define in passing some of the notation used by Levinson et al. (e_i, d_{jk} and c_{ij}), as this will be useful in due course. By equation (2.5), the probability that the initial state is i, given the observations, is just

$$\begin{aligned} P(C_1 = i \mid S_1, \ldots, S_T) &= P(S_1, \ldots, S_T, C_1 = i)/L_T \\ &= \alpha_1(i)\beta_1(i)/L_T. \end{aligned} \quad (2.7)$$

This suggests that this last quantity, computed on the basis of the current estimates of the parameters, may provide an improved estimate of δ_i. (In the notation of Levinson *et al.* the 're-estimate' (2.7) is e_i/L_T, or equivalently $e_i/\sum_{i=1}^m e_i$.)

To motivate an estimate for π_{kj}, we note that, given the observations, the expected number of occurrences of state j is

$$\sum_{t=1}^T P(C_t = j \mid S_1, \ldots, S_T) = \sum_{t=1}^T \alpha_t(j)\beta_t(j)/L_T.$$

The expected number of such occurrences for which $S_t = k$ is

$$\sum_{t=1}^T P(S_t = k, C_t = j \mid S_1, \ldots, S_T)$$

$$= \sum_{\{t:s_t=k\}} P(C_t = j \mid S_1, \ldots, S_T)$$

$$= \sum_{\{t:s_t=k\}} \alpha_t(j)\beta_t(j)/L_T.$$

(In the notation of Levinson *et al.* this is d_{jk}/L_T.) The ratio

$$\frac{\sum_{\{t:s_t=k\}} \alpha_t(j)\beta_t(j)}{\sum_{t=1}^T \alpha_t(j)\beta_t(j)} = \frac{d_{jk}}{\sum_{k=1}^n d_{jk}} \quad (2.8)$$

is therefore the expected proportion of the occurrences of state j for which the corresponding observation is k. This expression, evaluated at the current parameter estimates, provides a new estimate of π_{kj}.

In similar fashion we see that the expected number of transitions out of state i, given the observations, is

$$\sum_{t=1}^{T-1} P(C_t = i \mid S_1, \ldots, S_T) = \sum_{t=1}^{T-1} \alpha_t(i)\beta_t(i)/L_T, \quad (2.9)$$

and the expected number from i to j is

$$\sum_{t=1}^{T-1} P(C_t = i, C_{t+1} = j \mid S_1, \ldots, S_T) \quad (2.10)$$

$$= \frac{1}{L_T} \sum_{t=1}^{T-1} P(S_1, \ldots, S_T \mid C_t = i, C_{t+1} = j) P(C_t = i) \gamma_{ij}.$$

But properties (2.2) and (2.4) imply that:

$$P(S_1, \ldots, S_T \mid C_t = i, C_{t+1} = j)$$
$$= P(S_1, \ldots, S_t \mid C_t = i) P(S_{t+1}, \ldots, S_T \mid C_{t+1} = j)$$

$$\begin{aligned}
&= \mathrm{P}(S_1,\ldots,S_t \mid C_t=i)\,\mathrm{P}(S_{t+1} \mid C_{t+1}=j) \\
&\qquad \times \mathrm{P}(S_{t+2},\ldots,S_T \mid C_{t+1}=j) \\
&= (\alpha_t(i)/\mathrm{P}(C_t=i))\,\pi_{s_{t+1}j}\,\beta_{t+1}(j).
\end{aligned}$$

Hence, finally, the expected number of transitions from i to j is

$$\gamma_{ij} \sum_{t=1}^{T-1} \alpha_t(i)\pi_{s_{t+1}j}\beta_{t+1}(j)/L_T.$$

(In the notation of Levinson et al. this is c_{ij}/L_T.) The resulting new estimate of γ_{ij} is the following ratio, evaluated at the current parameter estimates:

$$\frac{\gamma_{ij}\sum_{t=1}^{T-1}\alpha_t(i)\pi_{s_{t+1}j}\beta_{t+1}(j)}{\sum_{t=1}^{T-1}\alpha_t(i)\beta_t(i)} = \frac{c_{ij}}{\sum_{j=1}^{m}c_{ij}}. \qquad (2.11)$$

It turns out that the re-estimates (2.7), (2.8) and (2.11) strictly increase the value of L_T, the likelihood function, except at critical points of L_T. We describe in outline how this may be established.

The concavity of the log function implies that the following inequality holds for a certain function Q, which will be defined later:

$$\log\left(L_T(\bar{\lambda})/L_T(\lambda)\right) \geq \left(Q(\lambda,\bar{\lambda}) - Q(\lambda,\lambda)\right)/L_T(\lambda).$$

Here $L_T(\lambda)$ denotes the likelihood of the observations $s_1,\ldots,s_T$ evaluated at parameter values $\lambda = (\delta,\Pi,\Gamma)$, and similarly $L_T(\bar{\lambda})$ for $\bar{\lambda} = (\bar{\delta},\bar{\Pi},\bar{\Gamma})$. Hence replacing λ by any $\bar{\lambda}$ such that $Q(\lambda,\bar{\lambda}) > Q(\lambda,\lambda)$ will increase the likelihood. Levinson et al. show that (as a function of $\bar{\lambda}$) $Q(\lambda,\bar{\lambda})$ is maximized by setting $\bar{\lambda}$ equal to $\hat{\lambda} = \left(\hat{\delta},\hat{\Pi},\hat{\Gamma}\right)$, where:

$$\begin{aligned}
\hat{\delta}_i &= e_i/\sum_{i=1}^{m}e_i \\
\hat{\pi}_{kj} &= d_{jk}/\sum_{k=1}^{n}d_{jk} \\
\hat{\gamma}_{ij} &= c_{ij}/\sum_{j=1}^{m}c_{ij}.
\end{aligned}$$

These are precisely the Baum–Welch re-estimates given in equations (2.7), (2.8) and (2.11). Furthermore, as Baum et al. (1970) show under mild assumptions (see p. 166 of their paper), the strict inequality $L_T(\hat{\lambda}) > L_T(\lambda)$ holds except when λ is a critical point of the likelihood, in which case $\hat{\lambda} = \lambda$. The Baum–Welch algorithm therefore guarantees an improvement in the likelihood except at a critical point of the likelihood.

The function Q referred to above is defined for all λ and $\bar{\lambda}$ by:

$$Q(\lambda, \bar{\lambda}) = \sum_{i=1}^{m} e_i \log \bar{\delta}_i + \sum_{j=1}^{m} \sum_{k=1}^{n} d_{jk} \log \bar{\pi}_{kj} + \sum_{i=1}^{m} \sum_{j=1}^{m} c_{ij} \log \bar{\gamma}_{ij}.$$

This is the function maximized by taking $\bar{\lambda}$ to be $\hat{\lambda}$, the Baum–Welch re-estimates. Examination of this expression reveals that the Baum–Welch algorithm, because it proceeds by maximizing $Q(\lambda, \bar{\lambda})$ as a function of $\bar{\lambda}$, is an example of the EM algorithm (as described by Little and Rubin, 1987, p. 130). In the present context the 'missing data' are the states $i_1, \ldots, i_T$ occupied by the Markov chain, and the 'complete data' are $s_1, \ldots, s_T, i_1, \ldots, i_T$. The quantity $Q(\lambda, \bar{\lambda})/L_T(\lambda)$ is just the complete-data log-likelihood, evaluated on the basis of the current parameter estimates λ, with those functions of the missing data which appear in it replaced by their conditional expectations given the observations. These conditional expectations are: $e_i/L_T(\lambda)$, the expectation of the number of times (0 or 1) that $C_1 = i$; $d_{jk}/L_T(\lambda)$, the expected number of times the observation is k when the state is j; and $c_{ij}/L_T(\lambda)$, the expected number of transitions from state i to state j.

Because of underflow and other problems, it is not possible to implement the forward-backward and Baum–Welch algorithms exactly as described above. Levinson *et al.* indicate how these problems may be overcome in practice, e.g. the use of scaling to prevent underflow in the computation of the forward and backward probabilities. We shall discuss scaling in some detail in section 2.5.

In speech recognition and other applications (see e.g. Fredkin and Rice, 1992a, or Guttorp, 1995, p. 101), it is of interest to determine the states of the Markov chain that are most likely (under the fitted model) to have given rise to the observation sequence. In the context of speech recognition this is known as the 'decoding' problem: see Juang and Rabiner (1991). More specifically, 'localized decoding' of the state at time t refers to the determination of that state $\bar{\imath}_t$ which is (*a posteriori*) most likely, that is:

$$\bar{\imath}_t = \arg \max_{1 \leq i \leq m} P(C_t = i \mid S_1 = s_1, S_2 = s_2, \ldots, S_T = s_T).$$

In contrast, 'global decoding' refers to the determination of the sequence of states $\hat{\imath}_1, \hat{\imath}_2, \ldots, \hat{\imath}_T$ which maximizes the conditional probability:

$$P(C_1 = i_1, C_2 = i_2, \ldots, C_T = i_T \mid S_1 = s_1, S_2 = s_2, \ldots, S_T = s_T);$$

HIDDEN MARKOV TIME SERIES MODELS: DEFINITION

or equivalently, and more conveniently, the joint probability:

$$P(C_1=i_1, C_2=i_2, \ldots, C_T=i_T, S_1=s_1, S_2=s_2, \ldots, S_T=s_T)$$
$$= (\delta_{i_1} \gamma_{i_1 i_2} \gamma_{i_2 i_3} \cdots \gamma_{i_{T-1} i_T})(_1\pi_{s_1 i_1} \cdots {}_T\pi_{s_T i_T}).$$

Global decoding can be carried out by using a dynamic programming method known as the Viterbi algorithm, for instance as follows.

Define
$$\xi_{1i} = P(C_1=i, S_1=s_1)$$
and, for $t=2, 3, \ldots, T$:
$$\xi_{ti} = \max_{i_1, i_2, \ldots, i_{t-1}} P(C_1=i_1, C_2=i_2, \ldots, C_{t-1}=i_{t-1}, C_t=i,$$
$$S_1=s_1, S_2=s_2, \ldots, S_t=s_t).$$

It can then be shown that the probabilities ξ_{tj} satisfy the following recursion for $t=1, 2, \ldots, T-1$:

$$\xi_{t+1,j} = \left[\max_i (\xi_{ti} \gamma_{ij})\right] \pi_{s_{t+1} j}.$$

This provides an efficient means of computing the $T \times m$ matrix of values ξ_{tj}, as the computational effort is linear in T. The required sequence of states $\hat{i}_1, \hat{i}_2, \ldots, \hat{i}_T$ can then be determined recursively from

$$\hat{i}_T = \arg \max_{1 \leq i \leq m} \xi_{Ti}$$

and, for $t=T-1, T-2, \ldots, 1$:

$$\hat{i}_t = \arg \max_{1 \leq i \leq m} (\xi_{ti} \gamma(i, \hat{i}_{t+1})).$$

(Here we have, for the sake of readability, written the transition probability $\gamma_{i \hat{i}_{t+1}}$ as $\gamma(i, \hat{i}_{t+1})$.) We note that the above algorithm is equally applicable whether the Markov chain underlying the model is stationary or nonstationary, provided that in the latter case the initial distribution is specified.

2.3 Hidden Markov time series models: definition and notation

In this section we introduce the time series models which will be studied in detail in the rest of the chapter. The notation is essentially that of the preceding section, but for ease of reference we define it in full here.

Let $\{C_t : t \in \mathbf{N}\}$ be an irreducible homogeneous Markov chain on the state space $\{1, 2, \ldots, m\}$, with transition probability matrix Γ. That is, $\Gamma = (\gamma_{ij})$, where for all states i and j and times t:

$$\gamma_{ij} = \mathrm{P}(C_t = j \mid C_{t-1} = i).$$

By the irreducibility of $\{C_t\}$, there exists a unique, strictly positive, stationary distribution, which we shall denote by the vector $\delta = (\delta_1, \delta_2, \ldots, \delta_m)$. We shall suppose (unless otherwise indicated) that $\{C_t\}$ is stationary, so that δ is for all t the distribution of C_t. (In this respect, the stationarity of the Markov chain, the models we consider here differ from the speech-processing models described in section 2.2, and from those of Albert (1991) and those of Leroux and Puterman (1992). Another respect in which our models differ from those used in speech processing is that the state space of the observations is not here assumed to be finite in general.)

Now let the nonnegative integer-valued random process $\{S_t : t \in \mathbf{N}\}$ be such that, conditional on $C^{(T)} = \{C_t : t = 1, \ldots, T\}$, the random variables $\{S_t : t = 1, \ldots, T\}$ are mutually independent and, if $C_t = i$, S_t takes the value s with probability ${}_t\pi_{si}$. That is, for $t = 1, \ldots, T$, the distribution of S_t conditional on $C^{(T)}$ is given by

$$\mathrm{P}(S_t = s \mid C_t = i) = {}_t\pi_{si}.$$

We shall refer to the probabilities ${}_t\pi_{si}$ as the 'state-dependent probabilities'. When these probabilities ${}_t\pi_{si}$ do not depend on t, the subscript t will be omitted. The two cases we shall discuss in detail are: (i) the conditional distribution of S_t is Poisson; and (ii) the conditional distribution of S_t is binomial.

In case (i), suppose that, if $C_t = i$, S_t has a Poisson distribution with mean λ_i. That is, we let $\mathrm{E}(S_t \mid C_t)$, the conditional mean of S_t, be

$$\mu(t) = \sum_{i=1}^{m} \lambda_i Z_i(t),$$

where the random variable $Z_i(t)$ is the indicator of the event $\{C_t = i\}$. The state-dependent probabilities are then given for all nonnegative integers s by:

$$\pi_{si} = e^{-\lambda_i} \lambda_i^s / s!.$$

In case (ii), we suppose that, if $C_t = i$, S_t has a binomial distribution with parameters n_t (a known positive integer) and p_i. That is, the conditional binomial distribution of S_t has parameters n_t

and $p(t)$, where
$$p(t) = \sum_{i=1}^{m} p_i Z_i(t),$$
and $Z_i(t)$ is, as before, the indicator of the event $\{C_t = i\}$. Then for $s = 0, 1, \ldots, n_t$:
$$_t\pi_{si} = \binom{n_t}{s} p_i^s (1-p_i)^{n_t - s}.$$

We shall refer to the models $\{S_t\}$ thus defined as (basic) Poisson- and binomial-hidden Markov models. In each case there are m^2 parameters: m parameters λ_i or p_i, and $m^2 - m$ transition probabilities γ_{ij}, e.g. the off-diagonal elements of Γ, to specify the 'hidden Markov chain' $\{C_t\}$. We note, however, that the off-diagonal elements must satisfy, in addition to the obvious nonnegativity constraints, the m constraints $\sum_{j \neq i} \gamma_{ij} \leq 1$, one for each i. The only constraints on λ_i and p_i are $\lambda_i \geq 0$ and $0 \leq p_i \leq 1$.

A useful device for depicting the dependence structure of such a model is the conditional independence graph: see for instance Whittaker (1990, Chapter 3). In such a graph the absence of an edge between two vertices indicates that the two variables concerned are conditionally independent given the other variables. Figure 2.1 displays the independence of the observations $\{S_t\}$ given the states $\{C_t\}$ occupied by the Markov chain, as well as the conditional independence of C_{t-1} and C_{t+1} given C_t, i.e. the Markov property.

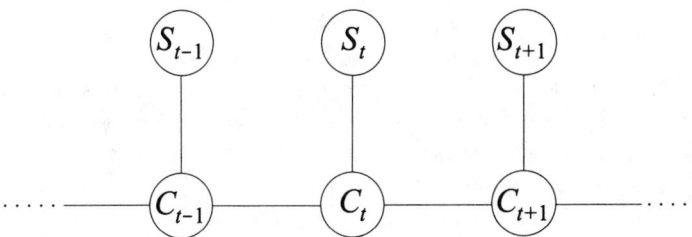

Figure 2.1. *Conditional independence graph of hidden Markov model.*

The special case $m = 2$ is of particular interest, as two-state models will often suffice in practice. If $m = 2$ we write the transition probability matrix of $\{C_t\}$ as

$$\Gamma = \begin{pmatrix} 1 - \gamma_1 & \gamma_1 \\ \gamma_2 & 1 - \gamma_2 \end{pmatrix},$$

and it follows that

$$\delta = \frac{1}{\gamma_1 + \gamma_2}(\gamma_2, \gamma_1). \qquad (2.12)$$

(Note that γ_1 and γ_2 are strictly positive by the irreducibility assumption, and that the two constraints $\sum_{j \neq i} \gamma_{ij} \leq 1$ reduce to $\gamma_i \leq 1$, $i = 1, 2$.) The following expression for Γ^k, obtained by diagonalizing Γ, will be useful in deriving the properties of $\{S_t\}$ when $m = 2$:

$$\Gamma^k = \begin{pmatrix} \delta_1 & \delta_2 \\ \delta_1 & \delta_2 \end{pmatrix} + w^k \begin{pmatrix} \delta_2 & -\delta_2 \\ -\delta_1 & \delta_1 \end{pmatrix}, \qquad (2.13)$$

where $w = 1 - \gamma_1 - \gamma_2$.

The case $m = 1$ is degenerate in the sense that a 'one-state hidden Markov model' is just a sequence of mutually independent random variables. As will be seen in Chapter 4, however, such a one-state model can be regarded as a useful baseline with which to compare a proposed model having two or more states.

We propose to show in this chapter and the next that the models $\{S_t\}$ defined above (and certain modifications and generalizations thereof) have properties which make them suitable as models for a wide range of discrete-valued time series. Although we have chosen to concentrate on the Poisson or binomial as the conditional distribution of the observations $\{S_t\}$, it will be clear that other discrete distributions, e.g. the negative binomial, may similarly be used. Use of the negative binomial in particular would allow us to introduce overdispersion via the conditional distribution of S_t given C_t, as well as via the parameter process $\{C_t\}$.

In fact a continuous conditional distribution can be used if a model for continuous observations is required. McInnes and Jack (1988, p. 32) report that such continuous-valued hidden Markov models have been applied more successfully in speech-recognition technology than discrete-valued models of the type described in the previous section. One other use of continuous-valued hidden Markov models is to generate models for point processes on the real line. That is, one takes the conditional distribution of the random variables S_t to be some continuous positive-valued distribution, and uses these random variables as the successive inter-event intervals in the point process. The resulting point process is a semi-

2.4 Correlation properties

An important characteristic of any time series model is its serial dependence structure. In the case of the usual normal-theory models this is specified by the autocorrelation function. While that is not true in general of nonnormal models, the autocorrelation function can still be a useful tool in the case of discrete-valued models, provided the observations are quantitative and not merely categorical. (Correlation is certainly not the only way one can seek to express serial dependence: an alternative which is useful for binary series has already been mentioned in section 1.2.2, namely the use of an odds ratio.)

The main purpose of this section is to derive expressions for the autocorrelations of the models defined in section 2.3, but in passing we discuss also various other properties such as means, variances and covariances, and show that a Poisson-hidden Markov model is overdispersed relative to the Poisson distribution.

As a preliminary step we state two results which will generally be of use in deriving moment properties for a hidden Markov model $\{S_t\}$. Firstly, provided the relevant expectations exist,

$$\mathrm{E}(f(S_t)) = \sum_{i=1}^{m} \mathrm{E}(f(S_t) \mid C_t = i)\, \delta_i. \quad (2.14)$$

This is proved by conditioning on C_t and noting that $\mathrm{P}(C_t = i) = \delta_i$. Secondly, provided again that the relevant expectations exist, we have for $k \in \mathbf{N}$ that

$$\mathrm{E}(f(S_t, S_{t+k}))$$
$$= \sum_{i,j=1}^{m} \mathrm{E}(f(S_t, S_{t+k}) \mid C_t = i, C_{t+k} = j)\, \delta_i \gamma_{ij}(k), \quad (2.15)$$

where $\gamma_{ij}(k) = (\Gamma^k)_{ij}$. To prove this, we condition on $C^{(t+k)}$ and exploit the fact that the conditional expectation of $f(S_t, S_{t+k})$, given $C^{(t+k)}$, is the conditional expectation given only C_t and C_{t+k}. Summing $\mathrm{P}(C_1, \ldots, C_{t+k})$ over the states at all times other than t and $t+k$ gives $\mathrm{P}(C_t, C_{t+k})$, and we therefore get

$$\mathrm{E}(f(S_t, S_{t+k}))$$
$$= \sum_{i,j} \mathrm{E}(f(S_t, S_{t+k}) \mid C_t = i, C_{t+k} = j)\, \mathrm{P}(C_t = i, C_{t+k} = j).$$

The result (2.15) then follows, since $P(C_t=i, C_{t+k}=j) = \delta_i \gamma_{ij}(k)$.

2.4.1 The autocorrelation function of a Poisson-hidden Markov model

Let $\{S_t\}$ be a Poisson-hidden Markov model, as defined in section 2.3. By equation (2.14), the mean is given by

$$E(S_t) = \sum_{i=1}^{m} E(S_t \mid C_t=i)\, \delta_i = \sum_{i=1}^{m} \lambda_i \delta_i = \delta \lambda',$$

where the vector λ is defined as $(\lambda_1, \ldots, \lambda_m)$. To derive the variance we need $E(S_t^2)$:

$$E(S_t^2) = \sum_{i=1}^{m} E(S_t^2 \mid C_t=i)\, \delta_i = \sum_{i=1}^{m} (\lambda_i^2 + \lambda_i)\, \delta_i.$$

It then follows that

$$\begin{aligned}
\mathrm{Var}(S_t) &= \sum (\lambda_i^2 + \lambda_i)\, \delta_i - \left(\sum \lambda_i \delta_i\right)^2 \\
&= \lambda D \lambda' + \delta \lambda' - (\delta \lambda')^2,
\end{aligned}$$

with D denoting $\mathrm{diag}(\delta)$. Alternatively, defining Λ as $\mathrm{diag}(\lambda)$, we may write this result as

$$\mathrm{Var}(S_t) = \delta \Lambda \lambda' + \delta \lambda' - (\delta \lambda')^2.$$

To find the covariance we note that, for $k \in \mathbf{N}$ but not $k=0$,

$$E(S_t S_{t+k} \mid C_t=i, C_{t+k}=j) = \lambda_i \lambda_j.$$

Hence by equation (2.15)

$$E(S_t S_{t+k}) = \sum_{i,j=1}^{m} \lambda_i \lambda_j \delta_i \gamma_{ij}(k) = \delta \Lambda \Gamma^k \lambda'.$$

The covariance is therefore

$$\mathrm{Cov}(S_t, S_{t+k}) = \delta \Lambda \Gamma^k \lambda' - (\delta \lambda')^2$$

and the autocorrelation function (ACF) is

$$\begin{aligned}
\rho_k &= \mathrm{Corr}(S_t, S_{t+k}) \\
&= \frac{\mathrm{Cov}(S_t, S_{t+k})}{\mathrm{Var}(S_t)} \\
&= \frac{\delta \Lambda \Gamma^k \lambda' - (\delta \lambda')^2}{\delta \Lambda \lambda' + \delta \lambda' - (\delta \lambda')^2}.
\end{aligned}$$

CORRELATION PROPERTIES

This expression for ρ_k is valid for all $k \in \mathbf{N}$. It is interesting to note that ρ_k depends on k only through Γ^k.

An alternative proof of the above results for the mean, variance and ACF proceeds via the conditional mean (and variance) $\mu(t) = \sum_{i=1}^{m} \lambda_i Z_i(t)$. The main steps in the proof are:

$$\begin{align}
\mathrm{E}(S_t) &= \mathrm{E}(\mu(t)) \tag{2.16}\\
\mathrm{E}(S_t^2) &= \mathrm{E}(\mu(t)^2 + \mu(t)) \tag{2.17}\\
\mathrm{E}(S_t S_{t+k}) &= \mathrm{E}(\mu(t)\mu(t+k))\\
\mathrm{Cov}(S_t, S_{t+k}) &= \mathrm{Cov}(\mu(t), \mu(t+k)).
\end{align}$$

As before, the result for the covariance is valid for $k \in \mathbf{N}$, but not $k=0$, relying as it does on the conditional independence of S_t and S_{t+k}.

We can establish that S_t is overdispersed (relative to the Poisson) either by using the conditional variance formula

$$\mathrm{Var}(S_t) = \mathrm{E}(\mathrm{Var}(S_t|C_t)) + \mathrm{Var}(\mathrm{E}(S_t|C_t)),$$

or by using steps (2.16) and (2.17) above. In any case it follows that

$$\mathrm{Var}(S_t) = \mathrm{E}(S_t) + \mathrm{Var}(\mu(t)). \tag{2.18}$$

Hence S_t is overdispersed unless the conditional mean $\mu(t)$ degenerates to a constant (i.e. unless the conditional mean does not depend at all on the current state of the Markov chain). If the hidden Markov chain has only two states ($m=2$), equation (2.18) implies that

$$\mathrm{Var}(S_t) = \mathrm{E}(S_t) + \delta_1 \delta_2 (\lambda_2 - \lambda_1)^2.$$

If $m=2$ we can derive also a simple explicit expression for the autocorrelation. From the expression (2.13) for Γ^k it follows that, for $k \in \mathbf{N}$,

$$\begin{align}
\rho_k &= \frac{\delta_1 \delta_2 (\lambda_2 - \lambda_1)^2}{\delta_1 \delta_2 (\lambda_2 - \lambda_1)^2 + \delta \lambda'} w^k \tag{2.19}\\
&= A w^k,
\end{align}$$

where $w = 1 - \gamma_1 - \gamma_2$. This formula for ρ_k can alternatively be derived by applying equation (10) on p. 196 of Cox and Lewis (1966). By taking $S_1, S_2, \ldots$ to be the inter-event intervals in a point process on the line, one generates a (nonorderly) semi-Markov process with two types of interval. Equation (10) gives the correlation between intervals in such a process, i.e. the correlation between S_t

and S_{t+k}. (Although the result of Cox and Lewis apparently refers to the case of continuous-valued intervals, its derivation applies equally to the discrete case.) If $\lambda_1 \neq \lambda_2$, the autocorrelation can be written as

$$\rho_k = \left(1 + \frac{\delta \lambda'}{(\lambda_2 - \lambda_1)^2 \delta_1 \delta_2}\right)^{-1} w^k. \qquad (2.20)$$

Since if $m = 2$ the autocorrelation function for the Markov chain $\{C_t\}$ is w^k (see section 1.2.1), formulae (2.19) and (2.20) display clearly the effect of the extra level of randomness present in a hidden Markov model: the autocorrelations are reduced by the factor A, which lies between 0 and 1. If $\lambda_1 = \lambda_2$, A attains the bound of 0, and S_t and S_{t+k} are uncorrelated. This is to be expected: if, conditional on $\{C_t\}$, S_t and S_{t+k} are independent and have distributions unaffected by the current state of $\{C_t\}$, relaxing the conditioning on $\{C_t\}$ will not induce any correlation between S_t and S_{t+k}. The upper bound of 1 may be approached, for example, by fixing λ_1 and letting λ_2 approach infinity. These observations, and the fact that w can be arbitrarily close to 1 (or -1), suggest that a wide range of correlations can be attained by at least the two-state models under discussion. In fact, given ϵ and η lying strictly between 0 and 1, the four parameters $\gamma_1, \gamma_2, \lambda_1$ and λ_2 can be chosen in such a way that the autocorrelation function reduces to $(1-\eta)(1-\epsilon)^k$: one possibility is to take $\gamma_1 = \gamma_2 = \epsilon/2$ and

$$\lambda_2 = \lambda_1 + (\nu - 1)^{-1}\{1 + (4\lambda_1(\nu - 1) + 1)^{1/2}\},$$

where $\nu = (1-\eta)^{-1}$. (Note that $\nu > 1$.) Similarly, one can choose the four parameters in such a way that the autocorrelation function reduces to $(1-\eta)(-1+\epsilon)^k$. Hence any autocorrelation function of the form

$$Aw^k \quad (0 < A < 1, \; -1 < w < 1)$$

is possible in the case $m = 2$.

For general m we recall from section 1.2.1 that 1 is a simple eigenvalue of the transition probability matrix Γ, and Γ can always be written in Jordan canonical form as $\Gamma = U\Omega U^{-1}$, where U, U^{-1} and Ω are of the following forms: $U = (\mathbf{1}' \; R)$, $U^{-1} = \begin{pmatrix} \delta \\ W \end{pmatrix}$ and $\Omega = \begin{pmatrix} 1 & \mathbf{0} \\ \mathbf{0}' & \Psi \end{pmatrix}$. Hence

$$\Gamma^k = U\Omega^k U^{-1} = \mathbf{1}'\delta + R\Psi^k W$$

and, for $k \in \mathbf{N}$,

$$\begin{aligned}
\mathrm{Cov}(S_t, S_{t+k}) &= \delta\Lambda(\mathbf{1}'\delta + R\Psi^k W)\lambda' - (\delta\lambda')^2 \\
&= (\delta\Lambda R)\Psi^k(W\lambda').
\end{aligned} \quad (2.21)$$

The covariance, and therefore the ACF, are thus seen to involve powers up to the kth of the eigenvalues of Γ. If all the quantities λ_i happen to be equal, i.e. if the conditional mean of S_t is unaffected by the value of C_t, the covariance and correlation are zero. This may be demonstrated as follows. Since $\begin{pmatrix} \delta \\ W \end{pmatrix}(\mathbf{1}'\ R) = U^{-1}U = I$, it follows that $W\mathbf{1}' = \mathbf{0}'$. Hence, if λ is a multiple of $\mathbf{1}$, $W\lambda' = \mathbf{0}'$ and the expression (2.21) reduces to zero.

If Γ is in fact diagonalizable, we have $\Omega = \mathrm{diag}(1, \omega_2, \ldots, \omega_m)$, the columns of U are right eigenvectors of Γ and the rows of U^{-1} left eigenvectors. In this case the covariance of S_t and S_{t+k} is more simply expressed as

$$\begin{aligned}
\mathrm{Cov}(S_t, S_{t+k}) &= (\delta\Lambda U)\Omega^k(U^{-1}\lambda') - (\delta\lambda')^2 \\
&= c\Omega^k d' - c_1 d_1 \\
&= \sum_{i=2}^{m} c_i \omega_i^k d_i,
\end{aligned}$$

where we have defined $c = \delta\Lambda U$ and $d' = U^{-1}\lambda'$. (Note that $c_1 = \delta\Lambda\mathbf{1}' = \delta\lambda'$ and $d_1 = \delta\lambda'$.) This expression for the covariance is valid for all $k \in \mathbf{N}$, but not for $k = 0$. If Γ is diagonalizable, therefore, the ACF is a linear combination of the kth powers of $\omega_2, \ldots, \omega_m$. These quantities ω_i, which are not necessarily distinct, are the eigenvalues of Γ other than 1. In modulus they may equal 1, but assuming aperiodicity of $\{C_t\}$ would rule out that possibility and give us the strict inequality $|\omega_i| < 1$ for all $i \geq 2$.

We illustrate the case of Γ nondiagonalizable by the following example.

Example Let the Markov chain $\{C_t\}$ have transition probability matrix

$$\Gamma = \begin{pmatrix} 1/3 & 1/3 & 1/3 \\ 2/3 & 0 & 1/3 \\ 1/2 & 1/2 & 0 \end{pmatrix}.$$

This matrix is not diagonalizable, but a Jordan canonical form

$\Gamma = U\Omega U^{-1}$ is

$$\begin{pmatrix} 1 & -1 & -9 \\ 1 & -1 & 15 \\ 1 & 3 & 0 \end{pmatrix} \begin{pmatrix} 1 & 0 & 0 \\ 0 & -1/3 & 1 \\ 0 & 0 & -1/3 \end{pmatrix} \frac{1}{96} \begin{pmatrix} 45 & 27 & 24 \\ -15 & -9 & 24 \\ -4 & 4 & 0 \end{pmatrix}.$$

Note that the first row of U^{-1} is the stationary distribution, δ. In the notation used above:

$$R = \begin{pmatrix} -1 & -9 \\ -1 & 15 \\ 3 & 0 \end{pmatrix}, \quad W = \frac{1}{96}\begin{pmatrix} -15 & -9 & 24 \\ -4 & 4 & 0 \end{pmatrix}$$

and $\Psi = \begin{pmatrix} -1/3 & 1 \\ 0 & -1/3 \end{pmatrix}$. From equation (2.21) the covariance of S_t and S_{t+k} is $(\delta\Lambda R)\Psi^k(W\lambda')$. Since $\Psi = -\frac{1}{3}I + N$, where N is the nilpotent matrix $\begin{pmatrix} 0 & 1 \\ 0 & 0 \end{pmatrix}$, it follows that

$$\Psi^k = \begin{pmatrix} (-1/3)^k & k(-1/3)^{k-1} \\ 0 & (-1/3)^k \end{pmatrix}.$$

For $k \in \mathbb{N}$, $\rho_k = \mathrm{Corr}(S_t, S_{t+k})$ is, as a function of k, a linear combination of $(-1/3)^k$ and $k(-1/3)^{k-1}$. The explicit expression is as follows:

$$\frac{(-\frac{1}{3})^k\{3(-5\lambda_1-3\lambda_2+8\lambda_3)^2+180(\lambda_2-\lambda_1)^2\}+k(-\frac{1}{3})^{k-1}\{4(-5\lambda_1-3\lambda_2+8\lambda_3)(\lambda_2-\lambda_1)\}}{32\{15(\lambda_1^2+\lambda_1)+9(\lambda_2^2+\lambda_2)+8(\lambda_3^2+\lambda_3)\}-(15\lambda_1+9\lambda_2+8\lambda_3)^2}.$$

2.4.2 The autocorrelation function of a binomial-hidden Markov model

Let $\{S_t\}$ be a binomial-hidden Markov model. The conditional distribution of S_t is binomial with the parameters n_t and $p(t) = \sum_{i=1}^m p_i Z_i(t)$. The derivation of the mean, variance and covariance function is very similar to that in the Poisson case, and will be described fairly briefly. We shall assume throughout this section (2.4.2) that Γ is diagonalizable, since that is the case of most practical interest and the modifications necessary if Γ is nondiagonalizable are quite analogous to those already described in section 2.4.1. We use the notation $p = (p_1, p_2, \ldots, p_m)$, $P = \mathrm{diag}(p)$ and (as in section 2.4.1) $D = \mathrm{diag}(\delta)$.

From equation (2.14) we have:

$$\mathrm{E}(S_t) = \sum_{i=1}^m (n_t p_i)\delta_i = n_t \delta p'$$

and

$$E(S_t^2) = \sum_{i=1}^{m}(n_t p_i(1-p_i) + n_t^2 p_i^2)\delta_i$$
$$= n_t \delta p' + n_t(n_t-1)pDp'$$
$$= n_t \delta p' + n_t(n_t-1)\delta Pp'.$$

Hence

$$\text{Var}(S_t) = n_t(n_t-1)\delta Pp' + n_t \delta p' - n_t^2(\delta p')^2$$
$$= n_t^2 \left(\delta Pp' - (\delta p')^2\right) + n_t(\delta p' - \delta Pp').$$

From equation (2.15) we have, for $k \in \mathbf{N}$,

$$E(S_t S_{t+k}) = \sum_{i,j=1}^{m}(n_t p_i)(n_{t+k} p_j)\,\delta_i \gamma_{ij}(k) = n_t n_{t+k}\delta P\Gamma^k p'.$$

Hence the result for the covariance is

$$\text{Cov}(S_t, S_{t+k}) = n_t n_{t+k}\left(\delta P\Gamma^k p' - (\delta p')^2\right).$$

Alternatively we could have proceeded via the following steps:

$$E(S_t) = n_t E(p(t))$$
$$E(S_t^2) = E(n_t p(t)(1-p(t)) + n_t^2 p(t)^2)$$
$$E(S_t S_{t+k}) = n_t n_{t+k} E(p(t)p(t+k))$$
$$\text{Cov}(S_t, S_{t+k}) = n_t n_{t+k} \text{Cov}(p(t), p(t+k))$$

and arrived at the same conclusions.

The autocorrelation is therefore given by:

$$\text{Corr}(S_t, S_{t+k}) = \frac{n_t n_{t+k}\left(\delta P\Gamma^k p' - (\delta p')^2\right)}{(\text{Var}(S_t)\text{Var}(S_{t+k}))^{1/2}}.$$

Provided that $\delta Pp' - (\delta p')^2 = \text{Var}(p(t))$ is nonzero, we can define

$$\alpha = \frac{\delta p' - \delta Pp'}{\delta Pp' - (\delta p')^2}$$

and write the autocorrelation as

$$(1+\alpha/n_t)^{-1/2}(1+\alpha/n_{t+k})^{-1/2}\frac{\delta P\Gamma^k p' - (\delta p')^2}{\delta Pp' - (\delta p')^2}. \qquad (2.22)$$

Since it can be shown that $\delta p' \geq \delta Pp'$, it follows that α is nonnegative, and the factors $(1+\alpha/n_t)^{-1/2}$ and $(1+\alpha/n_{t+k})^{-1/2}$ are at most 1. If n_t is constant with respect to t, expression (2.22) shows that the autocorrelation depends on k only through Γ^k (as in the Poisson case).

It is of interest here also to find a simple expression for the autocorrelation in the case $m = 2$, i.e. if the Markov chain has only two states. In that case, with w denoting $1 - \gamma_1 - \gamma_2$ as usual,

$$\delta P p' - (\delta p')^2 = \delta_1 \delta_2 (p_2 - p_1)^2,$$

$$\delta P \Gamma^k p' - (\delta p')^2 = \delta_1 \delta_2 (p_2 - p_1)^2 w^k$$

and

$$\delta p' - \delta P p' = \delta_1 p_1 (1 - p_1) + \delta_2 p_2 (1 - p_2).$$

Hence the autocorrelation is just $(1+\alpha/n_t)^{-1/2}(1+\alpha/n_{t+k})^{-1/2} w^k$, where α is given by

$$\frac{\delta_1 p_1 (1 - p_1) + \delta_2 p_2 (1 - p_2)}{\delta_1 \delta_2 (p_2 - p_1)^2}.$$

If n_t is constant with respect to t, we have $\rho_k = (1 + \alpha/n)^{-1} w^k$, which could also have been deduced from equation (10) on p. 196 of Cox and Lewis (1966). Again we see the correlation-reducing effect of the extra level of randomness imposed on top of the Markov chain $\{C_t\}$. But if $p_1 = 0$ and $p_2 = 1$ (and n_t is general), $\alpha = 0$ and the autocorrelation is exactly w^k: the hidden Markov model collapses to a Markov chain. The other extreme is that $p_1 = p_2$: in that case the autocorrelation is zero. If $n_t = n$ for all t we can, as in the case of a Poisson-hidden Markov model, achieve an autocorrelation of $(1-\eta)(1-\epsilon)^k$ or $(1-\eta)(-1+\epsilon)^k$ for given η and ϵ lying strictly between 0 and 1. To do so we can, for instance, let γ_1 and γ_2 both equal $\epsilon/2$ or $1 - \epsilon/2$, and take $p_1 = 0$ and $p_2 = (1+n(\nu-1)/2)^{-1}$, where ν (> 1) denotes $(1-\eta)^{-1}$. The reduction factor $(1+\alpha/n)^{-1}$ is then $1-\eta$, as required, and $w = 1 - \epsilon$ or $-1 + \epsilon$. Hence any ACF of the form

$$A w^k \quad (0 < A \leq 1, -1 < w < 1)$$

is possible in the case of n_t constant and $m = 2$.

For general m and Γ diagonalizable we can express the covariance of S_t and S_{t+k} in terms of the kth powers of the eigenvalues ω_i (other than 1) of Γ. With U as in section 2.4.1, $c = \delta P U$ and $d' = U^{-1} p'$, we have for $k \in \mathbf{N}$

$$\text{Cov}(S_t, S_{t+k}) = n_t n_{t+k} \sum_{i=2}^{m} c_i \omega_i^k d_i,$$

and a corresponding expression for the autocorrelation can be written down.

The case $n_t = 1$ (and m general) merits some attention, as it provides models for binary time series and the results derived above simplify to the following:

$$\mathrm{E}(S_t) = \delta p'$$

$$\mathrm{Var}(S_t) = \delta p' - (\delta p')^2$$

$$\mathrm{Corr}(S_t, S_{t+k}) = \frac{\delta P \Gamma^k p' - (\delta p')^2}{\delta p' - (\delta p')^2}.$$

2.4.3 The partial autocorrelation function

As in the case of a stationary Markov chain (see section 1.2.1), or for that matter any second-order stationary process, it is possible to deduce the partial autocorrelations ϕ_{kk} of a (stationary) binomial- or Poisson-hidden Markov model from the autocorrelations ρ_k by means of $\phi_{kk} = |P_k^*|/|P_k|$. The case of Markov chains warns us, however, that we should not expect any cut-off property except perhaps in hidden Markov models based on a two-state Markov chain. Even in that case, since ρ_k is of the form Aw^k for all $k \geq 1$, we find that $|P_2^*| = \rho_2 - \rho_1^2 = Aw^2 - (Aw)^2$: apart from some degenerate cases, ϕ_{22} is nonzero. It seems, therefore, that there is no cut-off property which might be useful for model identification purposes.

2.5 Evaluation of the likelihood function

Consider a sequence of observations $s_1, s_2, \ldots, s_T$ assumed to be generated by a hidden Markov model $\{S_t\}$ of either Poisson or binomial type. For several reasons it is important that we should be able to evaluate routinely the joint probability

$$L_T = \mathrm{P}(S_1 = s_1, S_2 = s_2, \ldots, S_T = s_T),$$

which we shall refer to, somewhat loosely, as the likelihood function. Firstly, it gives the finite-dimensional distributions of the process $\{S_t\}$, hence all the marginal, joint and conditional probabilities associated with the random variables S_1, S_2, etc. Secondly, by maximizing L_T with respect to Γ and λ (or Γ and p, whichever is appropriate) we can estimate the m^2 parameters of the model. In parameter-driven processes, of which a hidden Markov model is an example, maximum likelihood estimation is often not possible:

see Azzalini (1982), for instance. It is therefore a very convenient property of the hidden Markov models we are discussing that the likelihood can be evaluated sufficiently fast to permit direct numerical maximization. We first derive an expression for the likelihood in terms of a multiple sum, and then show how this may be rewritten in matrix notation in a way which suggests an efficient algorithm for computing L_T.

Given $C^{(T)}$, the joint probability of $S_1, \ldots, S_T$ is, by the assumed conditional independence, the product of Poisson or binomial probabilities $_t\pi_{si}$. More specifically, if we condition on the event $\{C_1 = i_1, \ldots, C_T = i_T\}$, the probability that $S_t = s_t$ for $1 \leq t \leq T$ is $_1\pi_{s_1 i_1}\, _2\pi_{s_2 i_2} \cdots _T\pi_{s_T i_T}$. To relax the conditioning we multiply by

$$P(C_1 = i_1, \ldots, C_T = i_T) = \delta_{i_1} \gamma_{i_1 i_2} \gamma_{i_2 i_3} \cdots \gamma_{i_{T-1} i_T}$$

and sum for all indices i_t over $\{1, 2, \ldots, m\}$. The result is that

$$L_T = \sum_{i_1=1}^{m} \cdots \sum_{i_T=1}^{m} \left(_1\pi_{s_1 i_1} \cdots _T\pi_{s_T i_T} \right)$$
$$\times \left(\delta_{i_1} \gamma_{i_1 i_2} \gamma_{i_2 i_3} \cdots \gamma_{i_{T-1} i_T} \right). \qquad (2.23)$$

As it stands, this expression is of little or no computational use, because it has m^T terms and cannot be evaluated except for very small T. The following slight rearrangement, however, enables us to write L_T in a more useful form:

$$L_T = \sum \cdots \sum \delta_{i_1}\, _1\pi_{s_1 i_1} \gamma_{i_1 i_2}\, _2\pi_{s_2 i_2} \gamma_{i_2 i_3} \cdots \gamma_{i_{T-1} i_T}\, _T\pi_{s_T i_T}$$
$$= \delta\, _1\lambda(s_1)\, \Gamma\, _2\lambda(s_2)\, \Gamma \cdots \Gamma\, _T\lambda(s_T)\mathbf{1}', \qquad (2.24)$$

where the matrices $_t\lambda(s)$ are defined by

$$_t\lambda(s) = \mathrm{diag}\left(_t\pi_{s1},\ _t\pi_{s2},\ \ldots,\ _t\pi_{sm} \right).$$

The matrix expression (2.24) for the likelihood, or a very similar one, appears for instance in Levinson et al. (1983, p. 1040) and Zucchini and Guttorp (1991). (As noted earlier, the subscript t before a symbol is necessary only in the binomial case, and then only if n_t is not constant with respect to t.)

The likelihood can now be written as

$$L_T = a \left(\prod_{t=2}^{T} B_t \right) \mathbf{1}'$$

if we define a by $a = \delta_1 \lambda(s_1)$ (i.e. $a_j = \delta_{j1} \pi_{s_1 j}$), and B_t as follows:

$$B_t = \Gamma_t \lambda(s_t) = (\gamma_{ij\,t} \pi_{s_t j}).$$

Equivalently, since $\delta \Gamma = \delta$, the likelihood can be written as

$$L_T = \delta \left(\prod_{t=1}^{T} B_t \right) \mathbf{1}'.$$

To evaluate L_T we can (in principle) just postmultiply a successively by B_2, B_3, etc. and add the elements of the resulting vector: 'in principle' because there is a numerical complication, which will be discussed below. The computational effort involved in such an algorithm is linear in T (as opposed to worse than exponential in the case of formula (2.23)) and quadratic in m. It is essentially the same as the case $t = T$ of the forward-backward algorithm $L_T = \sum_i \alpha_t(i) \beta_t(i)$ described in section 2.2. The case $t = 1$ would correspond to beginning with $\mathbf{1}'$, and premultiplying it successively by $B_T, \ldots, B_2$ and a. Other choices of t would correspond to other ways yet of 'bracketing' the matrix product $\prod_{t=2}^{T} B_t$.

To evaluate the likelihood as above we need δ, the stationary distribution of the Markov chain. This is computed as follows. We require that solution of the indeterminate system of linear equations $\delta \Gamma = \delta$ which also satisfies the constraint $\sum \delta_i = 1$. To find this solution, we simply replace one of the equations in the system (e.g. the last equation) by $\sum \delta_i = 1$, and use a standard routine for solving linear equations. For the case of a two-state model there is the explicit expression for δ given in equation (2.12).

The numerical complication referred to is that the computation of L_T as described may suffer from underflow even for relatively small values of T: the elements of the vector of forward probabilities $\alpha_r = a \prod_{t=2}^{r} B_t$ held at a particular stage of the algorithm may be too small to be distinguishable from zero, even if the quantities δ_i and π_{si} are all of moderate size. Since the likelihood is additive in form, it is not possible merely to work with logarithms. What can be done, however, and was done in the applications described in Chapter 4, is to scale the vector α_r at each stage so that the average element is 1, i.e. to divide α_r by $m^{-1} \sum_{i=1}^{m} \alpha_r(i)$, and accumulate the logarithms of these scale factors. Once the scaled likelihood has been computed thus, the sum of the logs of the scale factors is added to the log of the scaled likelihood. This procedure will avoid underflow in many cases and will yield $\log L_T$. Clearly many variations of this technique are possible: the scale factor could be chosen instead to be the largest element of the vector α_r, or the sum of the elements. Levinson *et al.* (1983) describe the use of this last possibility for scaling the forward and backward probabilities and computing the log of the likelihood.

From the above discussion it will be apparent that we disagree with Albert (1991), who states that the evaluation of the likelihood (conditional on the first state occupied by the Markov chain) is computationally infeasible, even for an observation sequence of moderate length and the two-state Markov chain he uses in his models. In order to estimate parameters he therefore seeks alternatives to direct numerical maximization, and implements a variation of EM involving an approximation at the E-step. For a fuller discussion of Albert's approach, including a response by Albert describing circumstances in which his estimation technique may nevertheless be valuable, see Le, Leroux and Puterman (1992).

2.6 Distributional properties

Given the values of the parameters Γ and λ (or p) of a hidden Markov model $\{S_t\}$, we are able to find the likelihood

$$\begin{aligned} L_T &= \mathrm{P}(S_1=s_1, S_2=s_2, \ldots, S_T=s_T) \\ &= \delta_1 \lambda(s_1) \Gamma_2 \lambda(s_2) \Gamma \cdots {}_T\lambda(s_T) \mathbf{1}' \end{aligned} \qquad (2.25)$$

of an observation sequence $s_1, s_2, \ldots, s_T$. It is largely a routine matter, therefore, to find various distributions of interest associated with the random variables $\{S_t\}$. One of the purposes of this section is to present some of these distributions and discuss questions of statistical interest arising from them. The other purposes are to derive certain probabilities associated with the Markov chain $\{C_t\}$, conditional on observations $s_1, \ldots, s_T$, and to discuss run-length distributions for binary hidden Markov models.

2.6.1 Marginal, joint and conditional distributions of the observations

We note, firstly, that some generalizations or modifications of equation (2.25) are easily proved. By considering time points $t, t+1, \ldots, T$ (rather than $1, 2, \ldots, T$) one arrives at

$$\begin{aligned} &\mathrm{P}(S_t=s_t, S_{t+1}=s_{t+1}, \ldots, S_T=s_T) \\ &= \delta_t \lambda(s_t) \Gamma_{t+1} \lambda(s_{t+1}) \Gamma \cdots {}_T\lambda(s_T) \mathbf{1}'. \end{aligned} \qquad (2.26)$$

DISTRIBUTIONAL PROPERTIES

Another kind of modification is exemplified by the following two probabilities:

$$\begin{aligned}
P(S_1=s, S_3=u, S_7=v) &= \sum_{i,j,k} {}_1\pi_{si}\, {}_3\pi_{uj}\, {}_7\pi_{vk} P(C_1=i, C_3=j, C_7=k) \\
&= \sum_{i,j,k} {}_1\pi_{si}\, {}_3\pi_{uj}\, {}_7\pi_{vk} \delta_i \gamma_{ij}(2)\gamma_{jk}(4) \\
&= \delta_1 \lambda(s)\, \Gamma^2\, {}_3\lambda(u)\, \Gamma^4\, {}_7\lambda(v)\, \mathbf{1}';
\end{aligned}$$

and

$$\begin{aligned}
P(S_t=u, S_{t+k}=v) &= \sum_{i,j} {}_t\pi_{ui}\, {}_{t+k}\pi_{vj} \delta_i \gamma_{ij}(k) \\
&= \delta_t \lambda(u)\, \Gamma^k\, {}_{t+k}\lambda(v)\, \mathbf{1}'. \qquad (2.27)
\end{aligned}$$

This points to an advantage of hidden Markov models as practical statistical tools: the ease with which the often-awkward issue of missing data may be handled. Suppose for instance that only one observation, s_l, is missing at random from an observation sequence of length T. (We consider only the case of observations missing at random.) The likelihood of the observations $s_1, \ldots, s_{l-1}, s_{l+1}, \ldots, s_T$ is

$$\begin{aligned}
&P(S_1=s_1, \ldots, S_{l-1}=s_{l-1}, S_{l+1}=s_{l+1}, \ldots, S_T=s_T) \\
&= \delta_1 \lambda(s_1)\, \Gamma \cdots \Gamma\, {}_{l-1}\lambda(s_{l-1})\, \Gamma^2\, {}_{l+1}\lambda(s_{l+1})\, \Gamma \cdots \Gamma\, {}_T\lambda(s_T)\mathbf{1}'.
\end{aligned}$$
(2.28)

The only difference between expression (2.28) and the standard expression (2.25) for the likelihood of the full observation sequence is that the diagonal matrix ${}_l\lambda(s_l)$ has been replaced by the identity matrix. In evaluating (2.28) one therefore proceeds as usual except that the probabilities ${}_l\pi_{s_l i}$ are all taken to be one. More generally, if several observations are missing, one replaces all the corresponding matrices ${}_t\lambda(s_t)$ by the identity.

To establish the marginal distribution of S_t we can either condition on C_t, or use the case $T=t$ of equation (2.26) as follows:

$$P(S_t=s) = \delta_t \lambda(s)\, \mathbf{1}' = \sum_{i=1}^m \delta_i\, {}_t\pi_{si}.$$

Hence S_t has a compound Poisson or compound binomial distribution, the compounding distribution being a discrete one, essentially the stationary distribution of the Markov chain. Bivariate distributions are available from

$$P(S_t = u, S_{t+1} = v) = \delta\, _t\lambda(u)\, \Gamma\, _{t+1}\lambda(v)\, \mathbf{1}' = \sum_{i,j=1}^{m} \delta_i\, _t\pi_{ui}\gamma_{ij}\, _{t+1}\pi_{vj}$$

or, more generally, from the expression (2.27). In the case $m = 2$ of a Poisson-hidden Markov model we therefore have, for instance,

$$P(S_t = u, S_{t+1} = v) = \delta_1(1 - \gamma_1)\pi_{u1}\pi_{v1} + \delta_1\gamma_1\pi_{u1}\pi_{v2} + \delta_2\gamma_2\pi_{u2}\pi_{v1} + \delta_2(1 - \gamma_2)\pi_{u2}\pi_{v2},$$

where $\Gamma = \begin{pmatrix} 1 - \gamma_1 & \gamma_1 \\ \gamma_2 & 1 - \gamma_2 \end{pmatrix}$, $\delta = \frac{1}{\gamma_1 + \gamma_2}(\gamma_2, \gamma_1)$, and $\pi_{si} = e^{-\lambda_i}\lambda_i^s/s!$ for all nonnegative integers s and $i = 1, 2$. It is clear that trivariate and higher-order joint distributions may similarly be obtained.

Since joint distributions are available, conditional distributions follow easily. One conditional distribution of particular statistical interest is the one-step-ahead forecast distribution, i.e. the distribution of S_{T+1}, given $S_1, \ldots, S_T$. This is given by a ratio of likelihood values as follows:

$$P(S_{T+1} = s_{T+1} \mid S_1 = s_1, \ldots, S_T = s_T)$$
$$= L_{T+1}/L_T$$
$$= \frac{\delta_1\lambda(s_1)\Gamma \cdots _{T+1}\lambda(s_{T+1})\mathbf{1}'}{\delta_1\lambda(s_1)\Gamma \cdots _T\lambda(s_T)\mathbf{1}'}.$$

More generally, the k-step-ahead forecast distribution is given by

$$P(S_{T+k} = s_{T+k} \mid S_1 = s_1, \ldots, S_T = s_T)$$
$$= P(S_1 = s_1, \ldots, S_T = s_T, S_{T+k} = s_{T+k})/L_T$$
$$= \delta_1\lambda(s_1)\Gamma \cdots _T\lambda(s_T)\Gamma^k\, _{T+k}\lambda(s_{T+k})\mathbf{1}'/L_T.$$

A question of interest which may be answered by evaluating conditional probabilities is whether hidden Markov models are in general themselves Markov processes. While it may seem obvious that the answer is no, for the sake of completeness a simple non-pathological counterexample to the Markov property is presented here.

Example Consider the binomial-hidden Markov model with the matrix $\Gamma = \frac{1}{4}\begin{pmatrix} 2 & 2 \\ 1 & 3 \end{pmatrix}$, $n_t = 1$ for all t, and $p = (\frac{1}{2}, 1)$. It follows

DISTRIBUTIONAL PROPERTIES

that $\delta = \frac{1}{3}(1,2)$ and $\lambda(1) = \text{diag}(p)$. The probabilities we need are:

$$\begin{aligned}
\text{P}(S_2=1) &= \delta\lambda(1)\mathbf{1}' \\
&= 5/6; \\
\text{P}(S_1=S_2=1) &= \text{P}(S_2=S_3=1) \\
&= \delta\lambda(1)\Gamma\lambda(1)\mathbf{1}' \\
&= 17/24; \\
\text{P}(S_1=S_2=S_3=1) &= \delta\lambda(1)\Gamma\lambda(1)\Gamma\lambda(1)\mathbf{1}' \\
&= 29/48.
\end{aligned}$$

Hence $\text{P}(S_3=1 \mid S_2=1) = 17/20$ and $\text{P}(S_3=1 \mid S_2=1, S_1=1) = 29/34$, which contradicts the Markov property.

In section 2.4.1, which discussed the ACF of Poisson-hidden Markov models, it was proved that, if all the elements of λ are equal, the random variables S_t and S_{t+k} are uncorrelated. This suggests that they may be independent. Using the matrix expression for the likelihood, we can in fact show, for Poisson- and binomial-hidden Markov models, that the random variables $S_1, \ldots, S_T$ are in general mutually independent if λ_i (or p_i) is constant. For notational convenience we demonstrate this only for Poisson- or stationary binomial-hidden Markov models, but the proof for the binomial case with nonconstant n_t is entirely similar.

Suppose, therefore, that π_{si} is, for each s, constant with respect to i. Then for all s,

$$\lambda(s) = \text{diag}(\pi_{s1}, \ldots, \pi_{sm}) = \pi_{s1}I_m.$$

Hence

$$\begin{aligned}
\text{P}(S_1=s_1, \ldots, S_T=s_T) &= \delta\lambda(s_1)\Gamma\lambda(s_2)\Gamma\cdots\Gamma\lambda(s_T)\mathbf{1}' \\
&= \pi_{s_11}\pi_{s_21}\cdots\pi_{s_T1}\delta\Gamma^{T-1}\mathbf{1}' \\
&= \pi_{s_11}\pi_{s_21}\cdots\pi_{s_T1},
\end{aligned}$$

since $\Gamma\mathbf{1}' = \mathbf{1}'$ and $\delta\mathbf{1}' = 1$. By the multiplicative form of their joint probability, $S_1, \ldots, S_T$ are mutually independent and

$$\text{P}(S_t=s) = \pi_{s1} = \begin{cases} e^{-\lambda}\lambda^s/s! \\ \binom{n}{s}p^s(1-p)^{n-s}. \end{cases}$$

Not unexpectedly, therefore, $S_1, \ldots, S_T$ are in this case mutually independent Poisson or binomial random variables. Even if n_t is not constant, $S_1, \ldots, S_T$ are mutually independent, although not identically distributed. In that case

$$P(S_t = s) = \binom{n_t}{s} p^s (1-p)^{n_t - s}.$$

In addition to the above, and the essentially equivalent case of a one-state hidden Markov model, we may note a third case in which the hidden Markov model degenerates to a sequence of independent random variables. If all the rows of the transition probability matrix Γ are equal, the Markov chain itself is a sequence of independent random variables. The usual link between observations at different times, via the underlying Markov chain, is thereby broken, and the observations $\{S_t\}$ are not merely conditionally independent, but independent. For a two-state Markov chain the property that the rows of the transition probability matrix are equal reduces to the statement $\gamma_1 + \gamma_2 = 1$. This is not surprising in view of the fact that the autocorrelation sequence of any two-state hidden Markov model we have discussed includes a factor of $(1 - \gamma_1 - \gamma_2)^k$.

However, there is a difference between this third case and the others as regards the marginal distributions, i.e. the distribution of the observations $\{S_t\}$. Whereas in the other cases the distribution will just be Poisson or binomial, the distribution may in this case be a nontrivial mixture of several distinct (Poisson or binomial) distributions. If the rows of the t.p.m. of the Markov chain are equal, the hidden Markov model may therefore be said to degenerate to an independent mixture model, in the sense in which that term is used by Leroux and Puterman (1992).

2.6.2 The Markov chain conditioned on the observations

Conditional probabilities of the form $P(C_t = i \mid S_1 = s_1, \ldots, S_T = s_T)$ may now also be derived: these results are a slight generalization of the corresponding ones of Zucchini and Guttorp (1991). We exclude for the moment the binomial case with nonconstant n_t.

Consider first the case $t > T$, that of 'state prediction'. We denote by $A_{\bullet i}$ the ith column of a matrix A, i.e. a column vector, and by $A_{i \bullet}$ the ith row. Since

$$\begin{aligned}
&P(S_1 = s_1, \ldots, S_T = s_T, C_t = i) \\
&= \sum_{i_1} \cdots \sum_{i_T} \left(\pi_{s_1 i_1} \cdots \pi_{s_T i_T} \right) \left(\delta_{i_1} \gamma_{i_1 i_2} \cdots \gamma_{i_{T-1} i_T} \gamma_{i_T i}(t-T) \right) \\
&= \delta \lambda(s_1) \Gamma \lambda(s_2) \cdots \Gamma \lambda(s_T) (\Gamma^{t-T})_{\bullet i},
\end{aligned}$$

DISTRIBUTIONAL PROPERTIES

we have, for $t > T$,

$$P(C_t = i \mid S_1 = s_1, \ldots, S_T = s_T)$$
$$= \delta\lambda(s_1)\Gamma\lambda(s_2)\cdots\Gamma\lambda(s_T)(\Gamma^{t-T})_{\bullet i}/L_T.$$

(Here, as before, an expression of the form $\gamma_{ij}(k)$ denotes the (i,j) element of the matrix Γ^k.) The case $t = T$, that of 'filtering', is proved similarly:

$$P(S_1 = s_1, \ldots, S_T = s_T, C_T = i)$$
$$= \sum_{i_1}\cdots\sum_{i_{T-1}}(\pi_{s_1 i_1}\cdots\pi_{s_{T-1} i_{T-1}}\pi_{s_T i})(\delta_{i_1}\gamma_{i_1 i_2}\cdots\gamma_{i_{T-2} i_{T-1}}\gamma_{i_{T-1} i})$$
$$= \delta\lambda(s_1)\Gamma\lambda(s_2)\cdots\Gamma\lambda(s_{T-1})\Gamma_{\bullet i}\pi_{s_T i}.$$

Hence

$$P(C_T = i \mid S_1 = s_1, \ldots, S_T = s_T)$$
$$= \delta\lambda(s_1)\Gamma\lambda(s_2)\cdots\Gamma\lambda(s_{T-1})\Gamma_{\bullet i}\pi_{s_T i}/L_T. \qquad (2.29)$$

The case $1 \leq t < T$, that of 'smoothing', is conveniently split into the two cases $1 < t < T$ and $t = 1$. For $1 < t < T$,

$$P(S_1 = s_1, \ldots, S_T = s_T, C_t = i)$$
$$= \sum_{i_1}\cdots\sum_{i_{t-1}}\sum_{i_{t+1}}\cdots\sum_{i_T}(\pi_{s_1 i_1}\cdots\pi_{s_{t-1} i_{t-1}}\pi_{s_t i}\pi_{s_{t+1} i_{t+1}}\cdots\pi_{s_T i_T})$$
$$\times (\delta_{i_1}\gamma_{i_1 i_2}\cdots\gamma_{i_{t-1} i}\gamma_{i i_{t+1}}\cdots\gamma_{i_{T-1} i_T})$$
$$= \delta\lambda(s_1)\Gamma\cdots\lambda(s_{t-1})\Gamma_{\bullet i}\pi_{s_t i}\Gamma_{i\bullet}\lambda(s_{t+1})\Gamma\cdots\Gamma\lambda(s_T)\mathbf{1}'.$$

This last quantity, divided by L_T, therefore gives the required conditional probability $P(C_t = i \mid S_1 = s_1, \ldots, S_T = s_T)$ for the case $1 < t < T$. Finally, for $t = 1$, we have

$$P(S_1 = s_1, \ldots, S_T = s_T, C_1 = i)$$
$$= \sum_{i_2}\cdots\sum_{i_T}(\pi_{s_1 i}\pi_{s_2 i_2}\cdots\pi_{s_T i_T})(\delta_i\gamma_{i i_2}\cdots\gamma_{i_{T-1} i_T})$$
$$= \delta_i\pi_{s_1 i}\Gamma_{i\bullet}\lambda(s_2)\Gamma\lambda(s_3)\cdots\Gamma\lambda(s_T)\mathbf{1}',$$

and division by L_T gives $P(C_1 = i \mid S_1 = s_1, \ldots, S_T = s_T)$.

In all four cases above, all that needs to be done to include binomial-hidden Markov models with nonconstant n_t is to insert a subscript, indicating the time, before each appearance of the symbols π and λ.

For $1 \leq t \leq T$, an alternative approach to the computation of the probabilities $P(C_t = i \mid S_1 = s_1, \ldots, S_T = s_T)$ is provided by the forward and backward probabilities, $\alpha_t(i)$ and $\beta_t(i)$, of section 2.2. Equation (2.5) of that section tells us that, for $1 \leq t \leq T$,

$$\alpha_t(i)\beta_t(i) = P(S_1 = s_1, \ldots, S_T = s_T, C_t = i),$$

and equation (2.6) that $\sum_{i=1}^{m} \alpha_t(i)\beta_t(i) = L_T$, for $1 \leq t \leq T$. If, therefore, we successively compute $\alpha_1(i), \alpha_2(i), \ldots, \alpha_t(i)$ and $\beta_T(i), \beta_{T-1}(i), \ldots, \beta_t(i)$ for each i, we can find the conditional distribution of C_t, given $S_1, \ldots, S_T$, as $\alpha_t(i)\beta_t(i)/L_T$. While it is true that the formulation of section 2.2 does not allow for the variation of the probabilities π_{si} with time, that is a modification easily introduced.

Finally we record here the joint distribution of $C_1, C_2, \ldots, C_T$ given the observations $s_1, s_2, \ldots, s_T$:

$$P(C_1 = i_1, \ldots, C_T = i_T \mid S_1 = s_1, \ldots, S_T = s_T)$$
$$= (_1\pi_{s_1 i_1} \cdots {_T\pi_{s_T i_T}})(\delta_{i_1} \gamma_{i_1 i_2} \cdots \gamma_{i_{T-1} i_T})/L_T.$$

In principle the corresponding conditional distribution for C_t only could be found from this expression by summation, but that is practicable only for very small T.

2.6.3 Runlength distributions for binary hidden Markov models

In certain applications of binary time series models, runlengths are of particular interest. For example, the distribution of runs of successive wet, or dry, days is of importance in modelling the occurrence of daily rainfall. For a binary (i.e. zero–one) Markov chain it is well known that the runlength is geometrically distributed: that is, if K denotes the length of a run of (e.g.) ones, then for all positive integers k and some $\lambda \in (0, 1)$:

$$P(K = k) = \lambda^{k-1}(1 - \lambda).$$

Here we give a general expression for the runlength distribution of a stationary binary hidden Markov model based on an m-state Markov chain. In some cases (but not all) this is a probabilistic mixture of several geometric distributions. There are also certain special cases (nontrivial in the sense that the model does not degenerate to a Markov chain) in which the runlength of ones is geometric.

Consider, therefore, a binary hidden Markov model $\{S_t\}$ in which the underlying stationary irreducible Markov chain has the states $1, 2, \ldots, m$, transition probability matrix Γ and stationary distribution δ. The probability of an observation being 1 in state i is

DISTRIBUTIONAL PROPERTIES

denoted by p_i. Several different definitions of runlength appear in the literature: the definition that we adopt is as follows. A run of ones is initiated by the sequence 01, and is said to be of length $k \in \mathbf{N}$ if that sequence is followed by a further $k - 1$ ones and a zero (in that order). A run of zeros is defined analogously. With K denoting the length of a run of ones, we therefore see that $P(K=k)$ is given by the following conditional probability:

$$\begin{aligned}
&P(K=k) \\
&= P(S_{t+2}=S_{t+3}=\ldots=S_{t+k}=1, S_{t+k+1}=0 \mid S_t=0, S_{t+1}=1) \\
&= \frac{\text{likelihood of sequence } 0\overbrace{11\ldots 1}^{k \text{ ones}}0}{\text{likelihood of sequence } 01} \\
&= \frac{\delta\lambda(0)\Gamma\lambda(1)\ldots\Gamma\lambda(1)\Gamma\lambda(0)\mathbf{1}'}{\delta\lambda(0)\Gamma\lambda(1)\mathbf{1}'} \\
&= \frac{\delta\lambda(0)B^k\Gamma\lambda(0)\mathbf{1}'}{\delta\lambda(0)B\mathbf{1}'},
\end{aligned}$$

where $\lambda(1) = \mathrm{diag}(p_1,\ldots,p_m)$, $\lambda(0) = I - \lambda(1)$, and B is defined as $\Gamma\lambda(1)$. If B has distinct eigenvalues λ_i, we can write B^k in diagonalized form to demonstrate the nature of its dependence on k:

$$B^k = U \,\mathrm{diag}(\lambda_1^k, \lambda_2^k, \ldots, \lambda_m^k)\, U^{-1}.$$

The columns of U are right eigenvectors corresponding to the eigenvalues λ_i. The required probability $P(K=k)$ is, as a function of k, a linear combination of the kth powers of these eigenvalues, and hence of $\lambda_i^{k-1}(1 - \lambda_i)$, $i = 1,\ldots,m$. In some cases, e.g. the first example discussed below, this yields an interpretation of K as a probabilistic mixture of geometric random variables. As the eigenvalues need not lie strictly between zero and one and can certainly be negative, such an interpretation is not always possible, and we shall not pursue it here.

Although it can be seen that slightly more general results are possible, we now restrict our attention to two-state models, with the t.p.m. given by

$$\Gamma = \begin{pmatrix} 1-\gamma_1 & \gamma_1 \\ \gamma_2 & 1-\gamma_2 \end{pmatrix}.$$

Here we see that, if one of the two eigenvalues (λ_2, say) is zero, $P(K=k)$ is, as a function of k, just proportional to λ_1^k. The case

$p_1 = 0$ and $p_2 \in (0,1)$ is a case in point: the matrix B is

$$\begin{pmatrix} 0 & \gamma_1 p_2 \\ 0 & (1-\gamma_2)p_2 \end{pmatrix},$$

and its eigenvalues are $\lambda_1 = (1-\gamma_2)p_2$ and $\lambda_2 = 0$. Since K is (with probability one) finite, the constant of proportionality is $(1-\lambda_1)/\lambda_1$, and the distribution of K is as follows, for all positive integers k:

$$P(K=k) = ((1-\gamma_2)p_2)^{k-1}(1-(1-\gamma_2)p_2).$$

Although not itself a Markov chain, this hidden Markov model has a geometric distribution for the length of a run of ones; for such a model the length of a run of zeros will not in general be geometric, however. It is also clear from probabilistic arguments why K is in general geometric if $p_1 = 0$, and the length of a run of zeros is not. For instance, if $S_{t+1} = 1$ (and $S_t = 0$ or 1), the Markov chain is necessarily in state 2 at time $t+1$. The next observation, S_{t+2}, will be 1 with probability $(1-\gamma_2) \times p_2$, regardless of the value of S_t.

We now show by way of examples that it is possible for a two-state binary hidden Markov model, as described above, to have mean runlength (of ones) either shorter or longer than that of a comparable Markov chain: comparable in the sense that (i) the unconditional probability of an observation being 1 is the same in all cases; and (ii) all autocorrelations are positive. Table 2.1 summarizes the properties of the models considered in these examples.

Examples

(a) Consider a binary hidden Markov model $\{S_t\}$ based on a two-state Markov chain with t.p.m.

$$\Gamma = \begin{pmatrix} 0.99 & 0.01 \\ 0.08 & 0.92 \end{pmatrix}$$

and $p = (0.1, 1)$. Then the stationary distribution of the Markov chain is $\delta = \frac{1}{9}(8,1)$, and $P(S_t = 1) = \delta p' = 0.2$. The matrix B is given by

$$\begin{pmatrix} 0.099 & 0.01 \\ 0.008 & 0.92 \end{pmatrix},$$

and its eigenvalues are $\lambda_1 = 0.9201$ and $\lambda_2 = 0.09890$. The distribution of K, the length of a run of ones, is for all $k \in \mathbf{N}$:

$$\begin{aligned} P(K=k) &= 0.009014\,\lambda_1^k + 8.165\,\lambda_2^k \\ &= 0.1038\lambda_1^{k-1}(1-\lambda_1) + 0.8962\lambda_2^{k-1}(1-\lambda_2). \end{aligned}$$

DISTRIBUTIONAL PROPERTIES

In this case, therefore, K is interpretable as a probabilistic mixture of geometric random variables. The mean of K is 2.294: see Table 2.1 for other properties. (The ACF of the model follows from the results of section 2.4.2.)

(b) Consider a model similar to Example (a), but with t.p.m.

$$\Gamma = \begin{pmatrix} 0.98 & 0.02 \\ 0.07 & 0.93 \end{pmatrix}$$

and $p = (0, 0.9)$. The stationary distribution δ is $\frac{1}{9}(7, 2)$ and, as for Example (a), $P(S_t = 1) = 0.2$. Since $p_1 = 0$, we can apply the result given above for such cases and conclude that K has the following geometric distribution:

$$P(K=k) = 0.837^{k-1} \times 0.163.$$

The mean of K is 6.135.

(c) A stationary Markov chain on 0 and 1 which is comparable to Examples (a) and (b), in that it also satisfies $P(S_t = 1) = 0.2$ and has positive autocorrelations, is the chain with t.p.m. $\begin{pmatrix} 0.9 & 0.1 \\ 0.4 & 0.6 \end{pmatrix}$ and corresponding stationary distribution $(0.8, 0.2)$. The distribution of the length of a run of ones is geometric:

$$P(K=k) = 0.6^{k-1} \times 0.4,$$

and the mean is therefore 2.5.

Table 2.1. *Properties of three comparable stationary binary time series models. Examples (a) and (b) are hidden Markov models, (c) is a comparable Markov chain.*

model	$P(K=1)$	$P(K=2)$	$P(K\leq 10)$	$E(K)$	σ_K	ACF
(a)	0.8159	0.0875	0.9549	2.294	5.212	0.5×0.91^k
(b)	0.1630	0.1364	0.8312	6.135	5.613	0.875×0.91^k
(c)	0.4000	0.2400	0.9940	2.500	1.936	0.5^k

An aspect of Example (a) which may be worth noting is the high probability (0.8159) that K, the length of a run of ones, equals 1. It is interesting to see how this probability arises. If $S_1 = 0$ and $S_2 = 1$ (i.e. if a run of ones starts at time 2), the probability that the underlying Markov chain $\{C_t\}$ is in state 1 at time 2 is given by equation (2.29) and equals 0.9083. We know also that

$$P(S_3=0 \mid C_2=1) = 0.99 \times 0.9 + 0.01 \times 0 = 0.891$$

and

$$P(S_3=0 \mid C_2=2) = 0.08 \times 0.9 + 0.92 \times 0 = 0.072.$$

Hence

$$P(S_3=0 \mid S_1=0, S_2=1) = 0.9083 \times 0.891 + 0.0917 \times 0.072 = 0.8159,$$

as already concluded. We may summarize this as follows. If we observe the sequence $S_1 = 0$, $S_2 = 1$, it is very likely (probability 0.9083) that the Markov chain is in state 1 at time 2. If so, then it is very likely (probability 0.891) that the next observation (S_3) is 0. Hence there is high probability (more than 0.9083×0.891) that a run of ones has length exactly 1.

2.7 Parameter estimation

2.7.1 Computing maximum likelihood estimates

Since the log-likelihood function can be evaluated routinely, even for very long sequences of observations, it is feasible to perform parameter estimation in hidden Markov models by direct numerical maximization of the log-likelihood.

We consider first the case of a Poisson-hidden Markov model in which the Markov chain has only two states, i.e. $m = 2$. The four parameters are, in the usual notation, $\gamma_1, \gamma_2, \lambda_1$ and λ_2. The only constraints on these parameters are $0 < \gamma_i \leq 1$ and $\lambda_i \geq 0$. For practical purposes these may be treated as $0 < \gamma_i < 1$ and $\lambda_i > 0$. The log-likelihood may then be reparametrized as the appropriate function of logit $\gamma_i = \log(\frac{\gamma_i}{1-\gamma_i})$ and $\log \lambda_i$, for $i = 1, 2$. An algorithm for unconstrained numerical maximization can be applied to obtain maximum likelihood estimates of logit γ_i and $\log \lambda_i$ ($i = 1, 2$) — or equivalently estimates of $\gamma_1, \gamma_2, \lambda_1$ and λ_2. A derivative-free algorithm such as the simplex algorithm of Nelder and Mead (Press et al., 1986, p. 289) is convenient for this purpose, although numerical differentiation of the likelihood would make possible the use of an algorithm requiring derivatives.

The case of a binomial-hidden Markov model with $m=2$ can be handled similarly. The four parameters are γ_1, γ_2, p_1 and p_2. The constraints on these parameters are $0 < \gamma_i \leq 1$ and $0 \leq p_i \leq 1$, but for practical purposes we can treat all four as lying strictly between 0 and 1, and apply the logistic transform in order to use an unconstrained maximization algorithm.

For $m > 2$ the generalized upper bound constraints already mentioned in section 2.3 must also be taken into account. That is, in maximizing the log-likelihood with respect to the $m^2 - m$ independent transition probabilities γ_{ij}, $i \neq j$, and the parameters λ_i or p_i, we must satisfy the m additional constraints $\sum_{j \neq i} \gamma_{ij} \leq 1$. This considerably alters the nature of the optimization problem: it is in this case necessary to maximize a (nonlinear) objective function subject to linear constraints other than simple lower and upper bounds of zero and one — preferably without supplying any derivatives of the objective. The program NPSOL (Gill et al., 1986) and the NAG version thereof, E04UCF (Numerical Algorithms Group, 1992), are designed to handle such problems (*inter alia*) and do not demand that values of the derivatives be supplied. The method used is a sequential quadratic programming algorithm. Although one can try to ensure that a global optimum is reached, by trying many sets of starting values of the parameters, there is no guarantee that this will succeed. This difficulty is discussed by Leroux and Puterman (1992), who present a systematic technique which generates many sets of starting values for their algorithm (described below) and thereby attempts to locate the global maximum.

In general it should be noted that the distribution of the observations is invariant under permutation of the states of the Markov chain, and this implies nonuniqueness of the maximum likelihood estimators. This is not in practice a problem, and one can if necessary order the states, e.g. in increasing order of λ_i or p_i. This can be done by adding the relevant constraints, e.g. $\lambda_1 \leq \lambda_2 \leq \ldots \leq \lambda_m$, to the optimization problem.

At this stage it is useful to compare the estimation problem described above, and its solution, with the estimation problem solved by the Baum–Welch algorithm. The two problems are by no means identical. Firstly, the speech-processing models of section 2.2 do not assume that the Markov chain is stationary, but here we do assume just that. Our probabilities δ_i are therefore not initial probabilities requiring estimation, but are the stationary probabilities completely determined by the transition probabilities γ_{ij}. Secondly, the speech-processing models involve an $n \times m$ matrix Π of probabilities, $(n-1)m$ of which are independently determined. We assume instead that the distribution of the observation for a given state (i) is dependent on one parameter only (λ_i or p_i), not $n-1$, and we allow the state space of the observations to be infinite. As it stands, therefore, the Baum–Welch algorithm is not applicable to the problem under discussion.

Another possible approach is contained in the work of Leroux and Puterman (1992), which deals *inter alia* with hidden Markov models with Poisson conditional distribution. Their models are intermediate between those discussed here and the speech-processing models. In their models the distribution of an observation given the current state of the Markov chain depends, as it does here, on only one parameter (λ_j in their notation), but the initial probabilities are not assumed to be the stationary probabilities. Leroux and Puterman maximize the likelihood with respect to the $m^2 - m$ independent transition probabilities, the m parameters λ_j, and the initial probabilities (but still take the number of parameters estimated to be m^2, e.g. in model selection). As they point out, that maximization can be accomplished by solving the m separate lower-dimensional maximization problems defined by starting from a fixed initial state: that is, one can take the initial probability of each of the states of the Markov chain in turn to be 1, and then choose as the initial state that one which produces the largest maximized log-likelihood. Although this property has been noted in the speech-processing literature (Levinson *et al.*, 1983, p. 1055), it does not seem to have been utilized explicitly except by Leroux and Puterman.

Leroux and Puterman use the EM algorithm to find the maximizing values of the transition probabilities and of the parameters λ_j. The 'missing data', i.e. the sequence of states $i_1, \ldots, i_T$ followed by the Markov chain, are represented by the indicator random variables defined as follows: $v_{jk}(t) = 1$ if $i_{t-1} = j$ and $i_t = k$, and $u_j(t) = 1$ if $i_t = j$. The complete-data log-likelihood is given by

$$\sum_{t=2}^{T} \log \gamma_{i_{t-1} i_t} + \sum_{t=1}^{T} \log \pi_{s_t i_t}$$

$$= \sum_{j=1}^{m} \sum_{k=1}^{m} \log \gamma_{jk} \sum_{t=2}^{T} v_{jk}(t) + \sum_{j=1}^{m} \sum_{t=1}^{T} u_j(t) \log \pi_{s_t j}.$$

This expression can be seen to consist of two parts: firstly, the log-likelihood of the Markov chain, conditioned on the initial state (i_1); and secondly, the log-likelihood of T independent observations. The first of these depends only on the parameters γ_{jk}, i.e. the transition probabilities, and the jth term of the second part (as on the right-

PARAMETER ESTIMATION

hand side above) depends only on the single parameter λ_j.

The E-step replaces $v_{jk}(t)$ and $u_j(t)$ in the complete-data log-likelihood by their conditional expectations given the observations (and current parameter estimates):

$$\hat{v}_{jk}(t) = \mathrm{P}(C_{t-1}=j, C_t=k \mid S_1, \ldots, S_T)$$

and

$$\hat{u}_j(t) = \mathrm{P}(C_t=j \mid S_1, \ldots, S_T).$$

The forward and backward probabilities, as defined in section 2.2, are needed to compute these conditional probabilities. Leroux and Puterman use the standard recursions (with scaling) to find the forward and backward probabilities, and derive the above conditional probabilities from them in the same way as described in section 2.2. The scaling method they use is to divide the forward probability $\alpha_t(i)$ by 10^p, where p is such that $10^{-p}\sum_i \alpha_t(i)$ lies between 0.1 and 1. The backward probabilities are scaled similarly.

The M-step separately maximizes the two parts of the complete-data log-likelihood. The Markov chain part is straightforward, since it is the standard problem of conditional maximum likelihood in a Markov chain, apart from the replacement of the missing data by their conditional expectations. The solution is otherwise as in section 1.2.1. The other part of the complete-data log-likelihood is maximized by setting λ_j equal to that value which maximizes

$$\sum_{t=1}^{T} \hat{u}_j(t) \log \pi_{s_t j}. \tag{2.30}$$

In the Poisson case this implies that

$$\lambda_j = \sum_{t=1}^{T} \hat{u}_j(t) s_t \bigg/ \sum_{t=1}^{T} \hat{u}_j(t).$$

Leroux and Puterman do not discuss models with binomial conditional distribution, i.e. the case

$$_t\pi_{sj} = \binom{n_t}{s} p_j^s (1-p_j)^{n_t-s},$$

but in that case the value of p_j which maximizes the expression (2.30) can be shown to be given by

$$p_j = \sum_{t=1}^{T} \hat{u}_j(t) s_t \bigg/ \sum_{t=1}^{T} \hat{u}_j(t) n_t.$$

Although neither the approach of Leroux and Puterman nor the Baum–Welch algorithm is directly applicable here, simply because the respective problems do not coincide, the EM algorithm can also be used for our models. The complete-data log-likelihood is in this case

$$\log \delta_{i_1} + \sum_{t=2}^{T} \log \gamma_{i_{t-1} i_t} + \sum_{t=1}^{T} \log \pi_{s_t i_t}$$

$$= \sum_{j=1}^{m} u_j(1) \log \delta_j + \sum_{j=1}^{m} \sum_{k=1}^{m} \left(\sum_{t=2}^{T} v_{jk}(t) \right) \log \gamma_{jk}$$

$$+ \sum_{j=1}^{m} \sum_{t=1}^{T} u_j(t) \log \pi_{s_t j},$$

with δ the stationary distribution implied by Γ. The simplest way of describing the method is to say that the estimation procedure of Leroux and Puterman is used except that the estimation of the transition probabilities γ_{jk} is based on the unconditional likelihood of a stationary Markov chain rather than on the likelihood of a (not necessarily stationary) Markov chain conditioned on the initial state. This does have the consequence that the neat explicit expression Leroux and Puterman can use to estimate the transition probabilities (their equation (6)) is replaced by an optimization problem of the following form, in the $m^2 - m$ off-diagonal transition probabilities γ_{jk}, $j \neq k$: subject to $\sum_{k \neq j} \gamma_{jk} \leq 1$ $(j=1,2,\ldots,m)$, and with δ denoting the stationary distribution implied by Γ, maximize $\sum_{j=1}^{m} a_j \log \delta_j + \sum_{j=1}^{m} \sum_{k=1}^{m} b_{jk} \log \gamma_{jk}$.

The numerical solution of such a problem may for instance be performed by NPSOL or E04UCF, with starting values of γ_{jk} supplied by the previous iteration of EM. It seems likely that such an EM algorithm, requiring the solution of a constrained nonlinear optimization problem at each M-step, would be slow compared to the direct numerical maximization technique described earlier in this section. It may be possible, however, to use a method for accelerating the EM algorithm, e.g. that of Meilijson (1989) or that of Jamshidian and Jennrich (1993).

Fredkin and Rice (1992b) discuss for their models the relative merits of the Baum–Welch algorithm and direct numerical maximization by means of NPSOL. They report that the use of NPSOL is much the more effective of the two approaches, as Baum–Welch can become bogged down by taking very small steps. For the sake of completeness we should mention that their interest is in hidden Markov models based on a continuous-time Markov chain, not necessarily stationary, which are sampled at discrete times. The observations are continuous-valued. Furthermore, the use of a continuous- rather than a discrete-time Markov chain in the specification of the model introduces a complication which makes the Baum–Welch algorithm awkward to use. The details are not relevant to our work, but the message does seem to be that the conceptual simplicity of direct numerical maximization makes adaptations or modifications of the models relatively easy to cope with.

2.7.2 Asymptotic properties of maximum likelihood estimators

The asymptotic properties of the maximum likelihood estimators in a hidden Markov model have been considered by (among others) Baum and Petrie (1966), Leroux (1992), Bickel and Ritov (1996), and Rydén (1994). Baum and Petrie prove results on consistency and asymptotic normality of these estimators for the case in which the hidden Markov model $\{S_t\}$ takes values in a finite state space. Leroux establishes consistency more generally (i.e. without the above finiteness assumption), and similarly Bickel and Ritov establish local asymptotic normality (as defined by Le Cam and Yang, 1990). However, as Rydén points out, this is insufficient to establish asymptotic normality, and the asymptotic normality of the ML estimators remains an open question. Rydén therefore proposes a new class of estimators, 'maximum split data likelihood estimators', and proves consistency and asymptotic normality of these, under suitable regularity conditions.

However, even if asymptotic normality of a class of estimators is established, there remain several problems if one wishes to use such results to construct confidence bounds for parameters. Firstly, the accuracy of the asymptotic approximation to the distribution of the estimators is unknown. Secondly, the results on asymptotic normality necessarily exclude cases in which the true value of a parameter is on the boundary of the parameter space. Hidden Markov models for which this is true of at least one of the parameters do seem to arise naturally in some applications: see for instance section 4.2.2.

An alternative is to use the parametric bootstrap, which we now describe.

2.7.3 Use of the parametric bootstrap

We illustrate the parametric bootstrap by showing how one can use it to estimate the variance-covariance matrix of the maximum-likelihood (or any other) estimators of the parameters of a two-state Poisson-hidden Markov model. The four parameters of the model are $\theta = (\gamma_1, \gamma_2, \lambda_1, \lambda_2)$, which determine the transition probability matrix

$$\Gamma = \begin{pmatrix} 1 - \gamma_1 & \gamma_1 \\ \gamma_2 & 1 - \gamma_2 \end{pmatrix},$$

and the corresponding stationary distribution

$$\delta = (\delta_1, \delta_2) = \frac{1}{\gamma_1 + \gamma_2}(\gamma_2, \gamma_1).$$

We wish to estimate the variance-covariance matrix of the estimator of θ, when the true parameter value is θ. We can estimate that matrix by the corresponding matrix for a model with true parameter value not θ, but our estimate from the data, $\hat{\theta}$. This latter variance-covariance matrix can in turn be approximated to any desired degree of accuracy by Monte Carlo methods using the algorithm described below.

First we simulate realizations of the fitted hidden Markov model, i.e. the model with parameters $\hat{\theta} = (\hat{\gamma}_1, \hat{\gamma}_2, \hat{\lambda}_1, \hat{\lambda}_2)$. We suppose that there is available a source of independent uniformly distributed random numbers, and also an independent source of independent Poisson-distributed random numbers. We generate an observation C_1^* from the stationary distribution $\hat{\delta}$ as follows: generate a uniform (0,1) random number U_1, and let

$$C_1^* = \begin{cases} 1 & \text{if } 0 \leq U_1 \leq \hat{\delta}_1 \\ 2 & \text{otherwise.} \end{cases}$$

Similarly, given $C_{t-1}^* = i$ for any $t = 2, 3, \ldots$, generate C_t^* from the probability distribution given by the ith row of the transition probability matrix $\hat{\Gamma}$. This provides a realization of the underlying Markov chain associated with t.p.m. $\hat{\Gamma}$. For each t we then let S_t^* be an independent Poisson random number with mean either $\hat{\lambda}_1$ or $\hat{\lambda}_2$, depending on whether C_t^* is 1 or 2.

The sequence $\{S_t^*\}$ is the required realization of the two-state Poisson-hidden Markov model associated with parameter vector $\hat{\theta}$.

We need to generate a large number, say B, of independent sequences $\{S_t^*\}$ and, for each one, compute the parameter estimates

IDENTIFICATION OF OUTLIERS

by the same method used to compute $\hat{\theta}$ from the original observations. (In our case this involves numerical maximization of the likelihood.) In this way we obtain B vectors, say $\hat{\theta}^*(1), \ldots, \hat{\theta}^*(B)$, one for each realization that we generate. The parametric bootstrap estimator for the variance-covariance matrix of $\hat{\theta}$ is then

$$\frac{1}{B-1} \sum_{b=1}^{B} \left(\hat{\theta}^*(b) - \hat{\theta}^*(\cdot)\right)' \left(\hat{\theta}^*(b) - \hat{\theta}^*(\cdot)\right),$$

where

$$\hat{\theta}^*(\cdot) = \frac{1}{B} \sum_{b=1}^{B} \hat{\theta}^*(b).$$

A detailed discussion of this and other bootstrap methods can be found in Efron and Tibshirani (1993). There it is indicated (p. 52) that $B = 50$ bootstrap replications are often enough to give a good estimate of a standard error, and that very seldom are more than 200 required.

Bias and confidence intervals for the individual parameters can also be estimated by means of the parametric bootstrap (Efron and Tibshirani, 1993, Chapters 10, 12–14).

Albert (1991) uses the parametric bootstrap to compute standard errors for his estimators, and we shall illustrate the technique in the examples presented in sections 4.2 and 4.8.

2.8 Identification of outliers

An important practical issue is whether, having fitted a hidden Markov (or other) model to an observed discrete-valued series s_1, s_2, ..., s_T, we can identify any of the observations as being outliers under the model chosen. Most of the literature on outliers in time series refers specifically to series modelled by Gaussian ARMA models: see for instance the work of Ljung (1993) and the references quoted therein. Since the available methods based on such ARMA models are not applicable to discrete-valued series, we describe here two straightforward but apparently new techniques for identifying outliers in (*inter alia*) discrete-valued series. These techniques are easy to use if the model is such that the relevant likelihood calculations can be performed routinely, which is certainly the case for hidden Markov models.

The first technique considers the observations one at a time and seeks those which, relative to the model and *all* the other observations in the series, are sufficiently extreme to suggest that they differ in nature or origin from the others. It relies on being able to compute $P(S_l = s \mid S_t = s_t, t \neq l)$ under the fitted model — i.e. the conditional (univariate) distribution of the lth observation, given all the other observations. This distribution is not difficult to find in the case of a hidden Markov model, and we describe the computation in more detail below. Having found this conditional distribution, we can then compute the conditional probabilities that $S_l \leq s_l$ and that $S_l \geq s_l$. If either of these probabilities is very small, the observation s_l may reasonably be regarded as an outlier relative to the model. In other words, the criterion suggested is that s_l is an outlier if it lies sufficiently far out in either tail of its conditional distribution.

The **p-value** of the lth observation, denoted by p_l, will be used to mean the smaller of

$$P(S_l \leq s_l \mid S_t = s_t, t \neq l) \qquad (2.31)$$

and

$$P(S_l \geq s_l \mid S_t = s_t, t \neq l). \qquad (2.32)$$

An excessively small value of p_l indicates that the observation s_l is an outlier. However, it is difficult to distinguish small probability values on a plot, and furthermore statisticians are accustomed to working with residuals rather than p-values when identifying outliers. We therefore suggest that the following **pseudo-residuals** r_l be plotted instead, in order to identify outliers:

$$r_l = \begin{cases} \Phi^{-1}(p_l) & \text{if } p_l \text{ is given by (2.31)} \\ \Phi^{-1}(1 - p_l) & \text{if } p_l \text{ is given by (2.32).} \end{cases}$$

Here Φ denotes the standard normal distribution function. An example of such a pseudo-residual plot appears in section 4.3.

For a hidden Markov model the computation of the required conditional probabilities proceeds as follows. We require, for each l from 1 to T, the probability that $S_l = s$, given that $S_t = s_t$ for all t other than l. It is sufficient to find this probability for each s from 0 to the observed value s_l. It is given by the ratio of the likelihood of all T observations to that of all observations but the lth. Since there is no difficulty in allowing for missing observations in computing likelihoods for hidden Markov models, this ratio is easily evaluated. In the notation of section 2.5, the probability is given explicitly for $l = 1, 2, \ldots, T$ by:

$$P(S_l = s \mid S_t = s_t, t \neq l)$$
$$= \frac{[\delta \Gamma_1 \lambda(s_1) \cdots \Gamma_{l-1} \lambda(s_{l-1})] \Gamma_l \lambda(s) [\Gamma_{l+1} \lambda(s_{l+1}) \cdots \Gamma_T \lambda(s_T) \mathbf{1}']}{[\delta \Gamma_1 \lambda(s_1) \cdots \Gamma_{l-1} \lambda(s_{l-1})] \Gamma I [\Gamma_{l+1} \lambda(s_{l+1}) \cdots \Gamma_T \lambda(s_T) \mathbf{1}']}$$
$$= \frac{(\delta B_1 B_2 \cdots B_{l-1}) \Gamma_l \lambda(s) (B_{l+1} \cdots B_T \mathbf{1}')}{(\delta B_1 B_2 \cdots B_{l-1}) \Gamma (B_{l+1} \cdots B_T \mathbf{1}')}.$$

In terms of the forward and backward probabilities of section 2.2, this is just:

$$\frac{\alpha_{l-1} \Gamma_l \lambda(s) \beta_l'}{\alpha_{l-1} \Gamma \beta_l'},$$

with α_0 being defined as δ.

In the case of the process $\{S_t\}$ being a stationary m-state Markov chain, with t.p.m. Γ and stationary distribution δ, the required conditional probability reduces to

$$P(S_l = s \mid S_t = s_t, t \neq l) = \frac{\gamma_{s_{l-1},s} \gamma_{s,s_{l+1}}}{\sum_{i=1}^m \gamma_{s_{l-1},i} \gamma_{i,s_{l+1}}}$$

for all l strictly between 1 and T. The remaining cases are as follows:

$$P(S_1 = s \mid S_2 = s_2, \ldots, S_T = s_T) = \frac{\delta_s \gamma_{s,s_2}}{\delta_{s_2}},$$

and

$$P(S_T = s \mid S_1 = s_1, \ldots, S_{T-1} = s_{T-1}) = \gamma_{s_{T-1},s}.$$

The second technique for outlier detection seeks observations which are extreme relative to the model and all *preceding* observations. In this case the relevant conditional probability is:

$$P(S_l = s \mid S_t = s_t, t < l).$$

This is given by the ratio of the likelihood of the first l observations to that of the first $l - 1$, which can be routinely evaluated for a hidden Markov model. For a Markov chain it is particularly simple, being just a one-step transition probability. We define the **forecast p-values** p_l^* and **forecast pseudo-residuals** r_l^* in the same way as above, except that the probabilities are conditioned on the observations preceding s_l rather than on all the observations other than s_l. That is, p_l^* is defined as the smaller of

$$P(S_l \leq s_l \mid S_t = s_t, t < l)$$

and

$$P(S_l \geq s_l \mid S_t = s_t, t < l),$$

and r_l^* as either $\Phi^{-1}(p_l^*)$ or $\Phi^{-1}(1-p_l^*)$, depending on whether the first or the second expression defines p_l^*. This second technique is applicable if one wishes to monitor a series on an ongoing basis, that is as each new observation becomes available. In effect one is examining how well each observation conforms to the corresponding forecast distribution. If an observation is either improbably small or improbably large under the forecast distribution, this indicates that the observation is an outlier or, alternatively, that the model, which was fitted on the basis of past observations, no longer provides an acceptable description of the series. Examples of the use of such forecast pseudo-residuals appear in sections 4.3 and 4.10.

A question of some theoretical and practical interest is whether, in the case of either of the above definitions of p-value, the p-values (and hence the corresponding pseudo-residuals) form an independent sequence. Even in the case of $\{S_t\}$ being a stationary two-state Markov chain, it can be shown that the 'lower-tail' forecast p-values are dependent. More specifically, if we define F_l by

$$F_l(s) = P(S_l \leq s \mid S^{(l-1)} = s^{(l-1)}) = P(S_l \leq s \mid S_{l-1} = s_{l-1}),$$

and similarly F_{l-1}, it can be shown that (apart from degenerate cases) $F_l(S_l)$ and $F_{l-1}(S_{l-1})$ are dependent. The same is true of (ordinary) p-values and pseudo-residuals. That is, if we define G_l by

$$G_l(s) = P(S_l \leq s \mid S_t = s_t, t \neq l),$$

and similarly G_{l-1}, it can be shown that $G_l(S_l)$ and $G_{l-1}(S_{l-1})$ are dependent. However, it seems doubtful that one can define useful residuals for discrete-valued time series that will turn out to be independent.

Such dependence makes it difficult to provide exact joint probability statements about either kind of p-value or pseudo-residual. However, a first-order Bonferroni bound can be used to make conservative statements of the following kind: the probability that all T observations in a series lie in the middle 99.99% of their respective conditional distributions is at least $1 - T/10\,000$. For T not too large this is close to the value based on an assumption of independence (0.9999^T). For percentages very close to 100 and T not too large, the error introduced by assuming independence will not be substantial, unless the series exhibits strong serial correlation.

In inspecting plots of pseudo-residuals, the reader should bear in mind that these are intended only as a convenient visual device for identifying outliers. The pseudo-residuals differ in certain important respects from the residuals obtained after fitting a model which assumes normality and independence of the observations.

Firstly, there is the dependence, already mentioned, among the pseudo-residuals, even if the model is assumed correct. Secondly, there are effects of the discreteness of the observations to be taken into account. For example, suppose the observed value s_l is 1, and a Poisson-hidden Markov model gives rise to the following distribution of S_l, conditional on the other observations (or on those preceding S_l): $P(S_l = 0) = 0.3$, $P(S_l = 1) = 0.3$, $P(S_l > 1) = 0.4$. The p-value, given by the lower tail, is 0.6, and because this exceeds 0.5, the corresponding pseudo-residual is positive. Cases such as this will not, however, give rise to apparent outliers and therefore do not present a problem. Thirdly, there may be effects of an asymmetry inherent in the nature of the observations. A series of unbounded counts with mean approximately 1 (say) is far more likely to include observations which are excessively large (by the criteria discussed here) than observations which are excessively small. This will be reflected in asymmetry in the corresponding pseudo-residuals.

Finally, it should also be noted that there are certain models, e.g. for binary series, that cannot give rise to outliers (in the senses used above) at all. Consider a binary hidden Markov model based on a two-state Markov chain, and with state-dependent probabilities 0.3 and 0.7 (say) of the observation being one. Whatever the neighbours or predecessors of an observation may be (and whatever the current state of the Markov chain), neither zero nor one is unlikely as an observation, and the p-values will not ever be sufficiently small to suggest that a particular observation is an outlier.

2.9 Reversibility

A random process is said to be reversible if its finite-dimensional distributions are invariant under reversal of time. More specifically, $\{X(t)\}$ is reversible if the random vector

$$(X(t_1), X(t_2), \ldots, X(t_n))$$

has the same distribution as

$$(X(\tau - t_1), X(\tau - t_2), \ldots, X(\tau - t_n))$$

for all positive integers n and all (appropriate) $\tau, t_1, \ldots, t_n$. In the case of a stationary irreducible Markov chain with transition probability matrix Γ and stationary distribution δ, it is necessary and sufficient for reversibility that the 'detailed balance conditions'

$$\delta_i \gamma_{ij} = \delta_j \gamma_{ji}$$

be satisfied for all states i and j (Kelly, 1979, p. 5). Equivalently, if the states are ordered in some way, it is necessary and sufficient that the detailed balance conditions be satisfied for all states i and j such that $i < j$. These conditions are trivially satisfied by all two-state stationary irreducible Markov chains, which are thereby reversible. The Markov chain of the example in section 2.4.1 is not reversible, however, because $\delta_1 \gamma_{12} = \frac{15}{32} \times \frac{1}{3} = \frac{5}{32}$ and $\delta_2 \gamma_{21} = \frac{9}{32} \times \frac{2}{3} = \frac{6}{32}$.

It is at least of probabilistic interest to know whether a given time series model is reversible. The classic example of a time series displaying irreversibility is (deseasonalized) streamflow. Since Gaussian processes are characterized by their first and second moments, the stationary normal-theory time series models are all reversible, and inappropriate as models for any series suspected of irreversibility. We show that, for $\{S_t\}$ a Poisson-hidden Markov or stationary binomial-hidden Markov model, reversibility of $\{C_t\}$ implies that of $\{S_t\}$, but not conversely: an irreversible $\{C_t\}$ may be associated with an $\{S_t\}$ either reversible or irreversible.

Let $\{C_t\}$, then, be reversible and let $\{S_t\}$ be as specified above. It will suffice to show that, for all T and all $s_1, s_2, \ldots, s_T$:

$$\begin{aligned} &P(S_1 = s_1, S_2 = s_2, \ldots, S_T = s_T) \\ &= P(S_T = s_1, S_{T-1} = s_2, \ldots, S_1 = s_T). \end{aligned}$$

One way to prove this is to use the matrix expression for the likelihood, i.e. to show that

$$\delta \lambda(s_1) \Gamma \lambda(s_2) \cdots \Gamma \lambda(s_T) \mathbf{1}' = \delta \lambda(s_T) \Gamma \lambda(s_{T-1}) \cdots \Gamma \lambda(s_1) \mathbf{1}'. \quad (2.33)$$

This can be done by writing Γ as AD, where as before D is defined as diag(δ), and the matrix A by $a_{ij} = \gamma_{ij}/\delta_j$. Note that, by the detailed balance conditions, A is symmetric. Note also that D and $\lambda(s)$, being diagonal matrices, commute under multiplication and are symmetric. The left-hand side of equation (2.33), being a scalar, equals its transpose, i.e. it equals:

$$\begin{aligned} &\mathbf{1} \lambda(s_T) DA \lambda(s_{T-1}) DA \cdots \lambda(s_2) DA \lambda(s_1)(\mathbf{1}D)' \\ &= \mathbf{1} D \lambda(s_T) AD \lambda(s_{T-1}) \cdots \lambda(s_2) AD \lambda(s_1) \mathbf{1}' \\ &= \delta \lambda(s_T) \Gamma \lambda(s_{T-1}) \cdots \lambda(s_2) \Gamma \lambda(s_1) \mathbf{1}'. \end{aligned}$$

This completes the proof. It is, however, interesting to note that another, perhaps more obvious, method of proof establishes this result without recourse to the Markov property of $\{C_t\}$: one simply conditions on $C^{(T)}$ and exploits the reversibility of $\{C_t\}$. The details are as follows:

$$\begin{aligned}
\mathrm{P}(S_1 = s_1, \ldots, S_T = s_T) \\
&= \sum_{i_1} \cdots \sum_{i_T} (\pi_{s_1 i_1} \pi_{s_2 i_2} \cdots \pi_{s_T i_T}) \, \mathrm{P}(C_1 = i_1, \ldots, C_T = i_T) \\
&= \sum_{i_1, \ldots, i_T} (\pi_{s_1 i_1} \cdots \pi_{s_T i_T}) \, \mathrm{P}(C_T = i_1, \ldots, C_1 = i_T) \\
&= \sum \mathrm{P}(S_T = s_1 \mid C_T = i_1) \cdots \mathrm{P}(S_1 = s_T \mid C_1 = i_T) \\
&\qquad \times \mathrm{P}(C_T = i_1, \ldots, C_1 = i_T) \\
&= \mathrm{P}(S_T = s_1, \ldots, S_1 = s_T).
\end{aligned}$$

(The second-last equality holds because $\mathrm{P}(S_t = s \mid C_t = i) = \pi_{si}$ for all t.)

To see that $\{C_t\}$ irreversible does not imply $\{S_t\}$ irreversible, let $\{C_t\}$ be irreversible and let $\{S_t\}$ be a Poisson-hidden Markov or stationary binomial-hidden Markov model based on it, with all the parameters λ_i or p_i equal. As demonstrated in section 2.6.1, $\{S_t\}$ is then just a sequence of independent and identically distributed random variables, and thereby reversible. At the end of this section we provide an example of a stationary irreversible hidden Markov model $\{S_t\}$, which is necessarily associated with an irreversible $\{C_t\}$.

One possible advantage of using a reversible rather than a general Markov chain in a hidden Markov model is parsimony. To specify a general chain on m states takes $m^2 - m$ parameters: to specify a reversible one takes $m - 1 + \binom{m}{2} = \frac{1}{2}(m-1)(m+2)$. This is because one needs to specify $m - 1$ of the elements of δ, and γ_{ij} for $i < j$. The remaining elements of Γ are then available from the detailed balance conditions and the row sum constraints on Γ. Hence there is a saving of

$$m^2 - m - (m-1)(m+2)/2 = (m-1)(m-2)/2$$

parameters in choosing $\{C_t\}$ to be reversible. As expected, there is no saving in the two-state case. One disadvantage of this approach, however, is that, if one seeks to maximize the likelihood with respect to the $\frac{1}{2}(m-1)(m+2)$ parameters, there are now nonlinear constraints of the form

$$\sum_{j=i+1}^{m} (\gamma_{ij} + \gamma_{ij}\delta_i/\delta_j) \leq 1$$

to be satisfied.

Given any stationary process $\{S_t\}$, a means of detecting irreversibility in some cases is to compare directional moments like $E(S_t S_{t+k}^2)$ and $E(S_t^2 S_{t+k})$. These will be equal if $\{S_t\}$ is reversible, otherwise possibly not. For that reason we include here a derivation of these quantities for both Poisson- and binomial-hidden Markov models. Although for the latter only the case of n_t constant is relevant to reversibility, the expressions for the directional moments are as easily derived for general n_t, and we do so. An economical way of proving all these results is to use equation (2.15) of section 2.4:

$$E(f(S_t, S_{t+k})) = \sum_{i,j=1}^{m} E\left(f(S_t, S_{t+k}) \mid C_t = i, C_{t+k} = j\right) \delta_i \gamma_{ij}(k).$$

The results for the Poisson case, valid for $k \in \mathbf{N}$, are then

$$E(S_t S_{t+k}^2) = \sum_{i,j=1}^{m} \delta_i \gamma_{ij}(k) \lambda_i (\lambda_j + \lambda_j^2)$$

and

$$E(S_t^2 S_{t+k}) = \sum_{i,j=1}^{m} \delta_i \gamma_{ij}(k) (\lambda_i + \lambda_i^2) \lambda_j.$$

For the binomial case they are

$$E(S_t S_{t+k}^2) = \sum_{i,j=1}^{m} \delta_i \gamma_{ij}(k) (n_t p_i) \left(n_{t+k} p_j (1 - p_j) + n_{t+k}^2 p_j^2\right)$$

and

$$E(S_t^2 S_{t+k}) = \sum_{i,j=1}^{m} \delta_i \gamma_{ij}(k) \left(n_t p_i (1 - p_i) + n_t^2 p_i^2\right) (n_{t+k} p_j).$$

With these expressions available for directional moments, we can now provide the promised example of a stationary irreversible hidden Markov model.

Example Let $\{C_t\}$ again be the (irreversible) Markov chain with transition probability matrix

$$\Gamma = \begin{pmatrix} 1/3 & 1/3 & 1/3 \\ 2/3 & 0 & 1/3 \\ 1/2 & 1/2 & 0 \end{pmatrix}$$

and stationary distribution $\delta = \frac{1}{32}(15, 9, 8)$. Let $\{S_t\}$ be defined as the binomial-hidden Markov model with $n_t = 2$ for all t and $p = (0, \frac{1}{2}, 1)$. Then the irreversibility of $\{S_t\}$ is shown by comparing $\mathrm{E}(S_t S_{t+1}^2)$ with $\mathrm{E}(S_t^2 S_{t+1})$:

$$\mathrm{E}(S_t S_{t+1}^2) = \sum_{i,j=1}^{3} \delta_i \gamma_{ij} \left(2p_i\right) \left(2p_j(1-p_j) + 4p_j^2\right) = 3/4;$$

$$\mathrm{E}(S_t^2 S_{t+1}) = \sum_{i,j=1}^{3} \delta_i \gamma_{ij} \left(2p_i(1-p_i) + 4p_i^2\right) (2p_j) = 25/32.$$

Alternatively, the irreversibility of $\{S_t\}$ may be established more directly by inspecting the joint probabilities $\mathrm{P}(S_t = u, S_{t+1} = v)$, which are shown below.

	$v = 0$	1	2
$u = 0$	31/128	5/64	7/32
1	3/32	0	3/64
2	13/64	1/16	7/128

2.10 Discussion

This chapter has introduced a class of hidden Markov models for time series of bounded or unbounded counts, that is for both 'binomial-like' and 'Poisson-like' observations. These models can accommodate a wider range of correlation structures than can Markov chains, or for that matter many of the other models described in Chapter 1. Negative correlation seems to be as easily modelled as positive, and the number of parameters is not excessive if m, the number of states of the Markov chain, is small. Furthermore, the relative ease with which parameters may be estimated by maximum likelihood contrasts quite sharply with the apparent difficulties of estimation in e.g. some of the discrete-valued ARMA processes of Chapter 1.

Katz (1981) discusses the relative merits of using Akaike's information criterion (AIC) and the Bayesian information criterion

(BIC) of Schwarz (1978) for estimating the order of a Markov chain. For several reasons he recommends the use of BIC for this purpose. In the case of hidden Markov models the problem of model selection (and in particular the choice of the number of states in the Markov chain component of the model) has yet to be satisfactorily solved. Of course it is simple enough to perform the computations required by AIC or BIC, and Titterington (1984) remarks in the context of independent mixture models that 'At the present ... there is no real formal alternative'. The use of AIC and BIC is illustrated in Chapter 4. It should be noted, however, that there are reservations about the theoretical justification for the use of such criteria even in the simpler problem of choosing the number of components in an independent mixture model: see Titterington (1984; 1990). In particular the regularity conditions under which the asymptotic properties of AIC are derived do not hold. Titterington (1990) lists this selection problem as a direction for future research. He also suggests an alternative approach, namely the use of Monte Carlo (bootstrap) methods. Although these methods are suggested in the context of independent mixtures, they are also applicable to hidden Markov models.

Rydén (1995) discusses the problem of estimating the order of a hidden Markov model, i.e. the number of states in the Markov chain. He shows that a certain class of estimators of this order, based on maximization of the split data likelihood, do not asymptotically underestimate the order. Included in this class are analogues of AIC and BIC, differing from AIC and BIC as we have defined them only in the use of the split data likelihood rather than the true likelihood.

The hidden Markov models discussed in this chapter are similar, but not identical, to models used in speech processing and the models of Leroux and Puterman, but the EM algorithm is more easily applied to these other classes of models than to the ones introduced here. Direct numerical maximization works well here, however, and EM is available as an alternative. An advantage of using models in which the Markov chain is assumed stationary is that the autocorrelation function is available as a tool for model verification (provided that n_t is constant if the model is one of binomial type). As will be seen in Chapter 3, however, the basic models of this chapter can be extended in various ways to cater for nonstationarity in the observations, without any transformation of the data being necessary, and without much modification of the estimation technique.

In some applications of hidden Markov models the states of the Markov chain may have, or may turn out to have, a useful substantive interpretation. That is, they may have a definite interpretation in terms of the subject-matter of the application concerned. For instance, the 'climate state' of the multisite precipitation models of Zucchini and Guttorp (1991) may correspond to a meteorologically defined weather state affecting all the sites. Even if the models are not substantive ones, however, they may be very useful as empirical ones. To illustrate this point we may consider the Gaussian ARMA models for continuous-valued time series: they are most often used as empirical models, i.e. without any close link to subject-matter considerations, yet they are no less valuable for that. (We use the terms 'empirical model' and 'substantive model' in the sense of Cox (1990).)

An important aspect of hidden Markov models which will now be discussed in detail is the ease with which the basic models of this chapter can be modified to accommodate a wide variety of different types of observation.

CHAPTER 3

Extensions and modifications

3.1 Introduction

One of the striking features of the hidden Markov time series models introduced in the previous chapter is the variety of ways in which they can be modified or generalized in order to provide models for a wider range of types of data. Since one can choose as the state-dependent distribution any discrete (or continuous) distribution, it is possible to model many different types of time series data by means of hidden Markov models. Chapter 2 introduced models for univariate series of bounded or unbounded counts. In this chapter we shall discuss models for several additional types of time series, including multivariate discrete-valued series, series of multinomial-like observations, categorical series, and series displaying time trend or seasonality or dependence on covariates other than time.

Most of the extensions we discuss involve modification of the state-dependent probabilities $_t\pi_{si}$, but in some cases it is more useful to modify $\{C_t\}$, the Markov chain component of the model. When seeking to incorporate seasonality into a model, for instance, one can do so either via the state-dependent distributions or via the transition probabilities of the Markov chain. The latter option can lead to more parsimonious models for multivariate time series exhibiting seasonal behaviour.

One minor modification of the parameter process, the use of a reversible Markov chain for $\{C_t\}$, has already been described in some detail in section 2.9. That modification resulted in a reduction in the number of parameters, provided that m, the number of states of $\{C_t\}$, exceeded two. More generally, any *a priori* restriction on the nature of the Markov chain, or equivalently on its transition probability matrix Γ, may result in a saving of parameters. For instance, if we could assume that $\gamma_{ij} = \gamma_{ji}$ for all states i and j, that would reduce the number of parameters in the model by

$\frac{1}{2}m(m-1)$. In all cases the use of the model would need to be justified by the *a priori* reasonableness of the restrictions placed on the structure of Γ. The evaluation of the likelihood could, however, proceed exactly as before, and the maximization thereof would present no new features apart from appropriate modifications of the constraints on the parameters.

The rest of this chapter will therefore discuss other extensions and modifications. Although most of these extensions fall into one of two broad classes (modifications of the parameter process and modifications of the state-dependent distributions), section 3.8 discusses a different kind of extension: the relaxation of the conditioning on the number of trials that is implicit in the use of binomial-hidden Markov models. That section introduces joint models for the number of trials and the number of successes in those trials. That is, the series $\{n_t\}$ is in that case also taken to be generated by some random process.

The purpose of this chapter is the exploration of statistically useful models and the derivation of the properties of those models. Accordingly we shall not always seek the greatest generality possible if that would not contribute to this purpose.

3.2 Models based on a second-order Markov chain

One fairly obvious extension of the basic hidden Markov models of Chapter 2 consists of replacing the underlying (first-order) Markov chain by a stationary second-order Markov chain $\{C_t\}$, on state space $M = \{1, 2, \ldots, m\}$. We suppose that $\{C_t\}$ has transition probabilities

$$p(i,j,k) = P(C_t = k \mid C_{t-1} = j, C_{t-2} = i)$$

and stationary bivariate distribution $u(j,k) = P(C_{t-1} = j, C_t = k)$. That is, the probabilities $u(j,k)$ satisfy

$$u(j,k) = \sum_{i=1}^{m} u(i,j) p(i,j,k)$$

and

$$\sum_{j=1}^{m} \sum_{k=1}^{m} u(j,k) = 1.$$

The process $\{C_t\}$ is not a (first-order) Markov chain, but if we make the definition $X_t = (C_{t-1}, C_t)$, then $\{X_t\}$ is indeed a Markov chain, on state space M^2. Given the three-dimensional array of probabilities $p(i,j,k)$, the matrix U of stationary bivariate probabilities $u(j,k)$ can be determined by finding the stationary distribution vector of $\{X_t\}$.

For instance, the most general stationary second-order Markov chain $\{C_t\}$ on the two states 1 and 2 can be characterized by the four transition probabilities:

$$a = P(C_t=2 \mid C_{t-1}=1, C_{t-2}=1),$$
$$b = P(C_t=1 \mid C_{t-1}=2, C_{t-2}=2),$$
$$c = P(C_t=1 \mid C_{t-1}=2, C_{t-2}=1),$$
$$d = P(C_t=2 \mid C_{t-1}=1, C_{t-2}=2).$$

The process $\{X_t\}$, as defined above, is then a first-order Markov chain, on the four states (1,1), (1,2), (2,1), (2,2) (in order), and with transition probability matrix:

$$\begin{pmatrix} 1-a & a & 0 & 0 \\ 0 & 0 & c & 1-c \\ 1-d & d & 0 & 0 \\ 0 & 0 & b & 1-b \end{pmatrix}.$$

The parameters a, b, c and d are bounded by 0 and 1 but are otherwise unconstrained. The stationary distribution of $\{X_t\}$ is proportional to

$$\bigl(\; b(1-d),\quad ab,\quad ab,\quad a(1-c) \;\bigr),$$

from which it follows that the matrix U of stationary bivariate probabilities for $\{C_t\}$ is

$$\frac{1}{b(1-d)+2ab+a(1-c)} \begin{pmatrix} b(1-d) & ab \\ ab & a(1-c) \end{pmatrix}.$$

A model $\{S_t\}$ based on a saturated second-order Markov chain $\{C_t\}$ on m states may be grossly overparametrized for statistical purposes. Using instead the Pegram or Raftery submodel (see section 1.3) can result in a considerable reduction in the number of parameters, as will be evident from Table 3.1. In that table it is assumed that the parameter process $\{C_t\}$ has m states, and that m parameters are needed to determine the state-dependent probabilities.

Among models with a second-order Markov chain as parameter process, the two-state case (as discussed above) is the most interesting, as it may provide a practical alternative to the use of an ordinary hidden Markov model with $m > 2$ states. That is, if a

Table 3.1. *A comparison of the numbers of parameters needed to specify various models of hidden Markov type. In all cases the parameter process has m states, and it is assumed that m parameters are needed to determine the state-dependent probabilities.*

parameter process $\{C_t\}$	no. of parameters to specify $\{C_t\}$	total no. to specify the model $\{S_t\}$
Markov chain	$m(m-1)$	m^2
general second-order Markov chain	$m^2(m-1)$	$m^3 - m^2 + m$
Pegram model for second-order MC	$m+1$	$2m+1$
Raftery model for second-order MC	$m(m-1)+1$	m^2+1

model based on a two-state first-order Markov chain is found to be inadequate, one can consider the use of a model based on a two-state second-order Markov chain instead of one based on (e.g.) a three-state first-order Markov chain. This last model would require a total of nine parameters, while the one based on the second-order chain would require either six or five, depending on whether a general second-order Markov chain or the Pegram–Raftery submodel is used as $\{C_t\}$. (Recall that for $m=2$ the Pegram and Raftery models are equivalent.)

We therefore elaborate here on the nature of a two-state Pegram–Raftery model of order two. The principal difference between such a model and a general two-state second-order Markov chain is the constraints placed on a, b, c and d. The Pegram–Raftery model requires that
$$c = l_1 b + l_2 (1-a) \tag{3.1}$$
and
$$d = l_1 a + l_2 (1-b), \tag{3.2}$$
where $l_1 + l_2 = 1$ but l_1 and l_2 are not in general assumed to be nonnegative. The three parameters of the Pegram–Raftery model can then be taken to be a, b and l_1, subject to c and d (as defined by equations (3.1) and (3.2)) being bounded by 0 and 1. If l_1 and

l_2 are in fact nonnegative, these bounds of 0 and 1 are redundant. If nonnegativity of l_1 and l_2 cannot be assumed, these (nonlinear) constraints will have to be imposed on the parameters a, b and l_1 when the likelihood is maximized.

We now consider the general problem of evaluating the likelihood of T consecutive observations from a model $\{S_t\}$ of hidden Markov type, but with the parameter process $\{C_t\}$ a second-order Markov chain on state space M, with transition probabilities $p(i,j,k)$ and stationary bivariate probabilities $u(j,k)$. We derive here an efficient recursive algorithm for computing such a likelihood.

For integers $t \geq 2$, define

$$\nu_t(i,j;s_1,\ldots,s_t) = P(S_1=s_1,\ldots,S_t=s_t,C_{t-1}=i,C_t=j).$$

For instance, $\nu_2(i,j;s_1,s_2) = {}_1\pi_{s_1 i}\, {}_2\pi_{s_2 j}\, u(i,j)$. It follows from this definition that

$$\nu_t(i_{t-1},i_t;s_1,\ldots,s_t)$$

$$= \sum_{i_1,i_2,\ldots,i_{t-2}=1}^{m} P(S_1=s_1,\ldots,S_t=s_t,C_1=i_1,\ldots,C_t=i_t)$$

$$= \sum_{i_1,i_2,\ldots,i_{t-2}} {}_1\pi_{s_1 i_1} \cdots {}_t\pi_{s_t i_t}$$
$$\times u(i_1,i_2)p(i_1,i_2,i_3)p(i_2,i_3,i_4)\cdots p(i_{t-2},i_{t-1},i_t).$$

This includes the case $t=2$, provided we interpret an empty product as 1 and note that for $t=2$ there is no summation. It then follows for integers $t \geq 3$ that:

$$\nu_t(i_{t-1},i_t;s_1,\ldots,s_t)$$

$$= {}_t\pi_{s_t i_t} \sum_{i_{t-2}} p(i_{t-2},i_{t-1},i_t) \left(\sum_{i_1,\ldots,i_{t-3}} {}_1\pi_{s_1 i_1} \cdots {}_{t-1}\pi_{s_{t-1} i_{t-1}} \right.$$

$$\left. \times u(i_1,i_2)p(i_1,i_2,i_3)\cdots p(i_{t-3},i_{t-2},i_{t-1}) \right)$$

$$= {}_t\pi_{s_t i_t} \sum_{i_{t-2}} p(i_{t-2},i_{t-1},i_t)\, \nu_{t-1}(i_{t-2},i_{t-1};s_1,\ldots,s_{t-1}). \quad (3.3)$$

Since we know $\nu_2(i,j;s_1,s_2)$ for all i and j, the above recursion enables us to compute $\nu_T(i,j;s_1,\ldots,s_T)$ for all i and j, and hence to find the required likelihood as follows:

$$P(S_1=s_1,\ldots,S_T=s_T) = \sum_i \sum_j \nu_T(i,j;s_1,\ldots,s_T). \quad (3.4)$$

The computational effort involved in this algorithm is linear in T and cubic in m. As in the case of the models based on a first-order Markov chain (see section 2.5), scaling is necessary in practice to avoid underflow. In that case a scaled version of the vector of forward probabilities had to be computed: here it is the $m \times m$ matrix of probabilities $\nu_t(i, j; s_1, \ldots, s_t)$ that must be scaled. It will be noted that the quantities $\nu_t(i, j; s_1, \ldots, s_t)$ are just an extension of the forward probabilities of section 2.2.

With the likelihood at our disposal, we are able to find the one-step-ahead forecast distribution exactly as before:

$$P(S_{T+1}=s_{T+1} \mid S_1=s_1, \ldots, S_T=s_T)$$
$$= \frac{P(S_1=s_1, \ldots, S_T=s_T, S_{T+1}=s_{T+1})}{P(S_1=s_1, \ldots, S_T=s_T)}.$$

The general k-step-ahead forecast distribution is slightly more awkward. If m^{k-1} is sufficiently small, we can evaluate

$$P(S_1=s_1, \ldots, S_T=s_T, S_{T+k}=s_{T+k})$$

as

$$\sum_{s_{T+1}, \ldots, s_{T+k-1}} P(S_1=s_1, \ldots, S_T=s_T,$$
$$S_{T+1}=s_{T+1}, \ldots, S_{T+k}=s_{T+k}). \quad (3.5)$$

Division by the likelihood of the first T observations then yields the required conditional probability, $P(S_{T+k}=s_{T+k} \mid S_1=s_1, \ldots, S_T=s_T)$.

It is interesting to note, however, that the recursive algorithm for finding ν_T and hence the likelihood function can be modified to provide an algorithm linear in k for computing the joint probability $P(S_1=s_1, \ldots, S_T=s_T, S_{T+k}=s)$. We proceed as follows: for integers $t \geq T+1$ and all states i and j, define

$$\phi_t(i, j; s_1, \ldots, s_T; s)$$
$$= P(S_1=s_1, \ldots, S_T=s_T, S_t=s, C_{t-1}=i, C_t=j).$$

It then follows that, for $t \geq T+2$,

$$\phi_t(i_{t-1}, i_t; s_1, \ldots, s_T; s)$$
$$= \sum_{i_1} \sum_{i_2} \cdots \sum_{i_{t-2}} {}_1\pi_{s_1 i_1} \cdots {}_T\pi_{s_T i_T} {}_t\pi_{s i_t}$$
$$\times u(i_1, i_2) p(i_1, i_2, i_3) \cdots p(i_{t-2}, i_{t-1}, i_t)$$

$$= \frac{{}_t\pi_{s i_t}}{{}_{t-1}\pi_{s i_{t-1}}} \sum_{i_{t-2}} p(i_{t-2}, i_{t-1}, i_t) \sum_{i_1,\ldots,i_{t-3}} {}_1\pi_{s_1 i_1} \cdots {}_T\pi_{s_T i_T}$$
$$\times \, {}_{t-1}\pi_{s i_{t-1}} u(i_1, i_2) p(i_1, i_2, i_3) \cdots p(i_{t-3}, i_{t-2}, i_{t-1})$$
$$= \frac{{}_t\pi_{s i_t}}{{}_{t-1}\pi_{s i_{t-1}}} \sum_{i_{t-2}} p(i_{t-2}, i_{t-1}, i_t) \phi_{t-1}(i_{t-2}, i_{t-1}; s_1, \ldots, s_T; s).$$

The recursion is started by noting that, for all i and j,
$$\phi_{T+1}(i, j; s_1, \ldots, s_T; s) = \nu_{T+1}(i, j; s_1, \ldots, s_T, s).$$
The joint probability we seek is then given by
$$P(S_1 = s_1, \ldots, S_T = s_T, S_{T+k} = s)$$
$$= \sum_i \sum_j \phi_{T+k}(i, j; s_1, \ldots, s_T; s). \quad (3.6)$$

Bivariate and marginal distributions for $\{S_t\}$ can also be written down:
$$P(S_t = u, S_{t+1} = v) = \sum_i \sum_j u(i, j) \, {}_t\pi_{ui} \, {}_{t+1}\pi_{vj},$$
and
$$P(S_t = u) = \sum_i \delta_i \, {}_t\pi_{ui},$$
where $\delta_i = P(C_t = i) = \sum_j u(i, j) = \sum_j u(j, i)$. One can use either (3.5) or (3.6) to find the joint probability $P(S_1 = u, S_{1+k} = v)$, and thereby the kth-order autocovariance and autocorrelation of a stationary $\{S_t\}$ based on a second-order Markov chain $\{C_t\}$. Models involving a binomial distribution with nonconstant n_t can be handled similarly, although (3.5) and (3.6) do not apply exactly as they stand, and would need to be generalized slightly to be relevant.

The reader interested in further work on hidden Markov models based on a higher-order Markov chain (in particular, those based on Raftery's mixture transition distribution model) is referred to the dissertation of Schimert (1992).

3.3 Multinomial-hidden Markov models

Here we consider the multinomial extension of binomial-hidden Markov models. Essentially all this involves is the use of q mutually exclusive categories (rather than merely two), into one of which the outcome of each trial falls. As in the binomial case, there are n_t trials at time t. One set of data for which such a model is useful is the series discussed in section 4.10, relating to homicides and suicides in Cape Town during the period from 1986 to 1991. There n_t represents the total number of deaths due to homicide or suicide in week t, and the $q=5$ categories are: firearm homicide, nonfirearm homicide, firearm suicide, nonfirearm suicide and 'legal intervention homicide'.

More formally, let $\{C_t\}$ be the usual stationary first-order Markov chain on m states, and suppose that, conditional on $C^{(T)}$, the T random vectors

$$\underline{S}_t = (S_{t1}, S_{t2}, \ldots, S_{tq}) \quad (t=1, 2, \ldots, T)$$

have independent multinomial distributions. We suppose in particular that, if $C_t = i$, then $(S_{t1}, S_{t2}, \ldots, S_{tq})$ has the multinomial distribution with parameters n_t (which is known) and $p_{i1}, p_{i2}, \ldots, p_{iq}$, with $\sum_{j=1}^{q} p_{ij} = 1$. There are therefore $m^2 - m + (q-1)m = m^2 + m(q-2)$ parameters. In addition to the usual constraints on the transition probabilities γ_{ij}, and the obvious nonnegativity requirements $p_{ij} \geq 0$, there are the m constraints $\sum_{j=1}^{q-1} p_{ij} \leq 1$, one for each state i of the Markov chain, on the $(q-1)m$ independently determined multinomial probabilities.

One way in which one might view such processes is that they provide models for time series of discrete compositional data. As most of the models for compositional data are based on continuous distributions (see e.g. Aitchison, 1986; or Grunwald, Raftery and Guttorp, 1993), this may be a useful perspective. The case $n_t = 1$, it will be noted, provides a model for a single categorical time series: at each time point there is one observation, which falls into one of the q categories. In section 3.3.3 we shall deal specifically with models of this kind.

3.3.1 The likelihood

The computation of the likelihood of observations $\underline{s}_1, \ldots, \underline{s}_T$ from a general multinomial-hidden Markov model differs little from the case of a binomial-hidden Markov model: the only difference is that the binomial probabilities

$$_t\pi_{s_t,i} = \binom{n_t}{s_t} p_i^{s_t}(1-p_i)^{n_t - s_t}$$

are replaced by the multinomial probabilities

$$_t\pi_{\underline{s}_t,i} = P(\underline{S}_t = \underline{s}_t \mid C_t = i) = \binom{n_t}{s_{t1}, s_{t2}, \ldots, s_{tq}} p_{i1}^{s_{t1}} p_{i2}^{s_{t2}} \cdots p_{iq}^{s_{tq}}.$$

Otherwise the computation proceeds as before. The likelihood is therefore given by

$$\sum_{i_1,\ldots,i_T=1}^{m} \left(\delta_{i_1}\gamma_{i_1 i_2}\cdots\gamma_{i_{T-1} i_T}\right)\left({}_1\pi_{\underline{s}_1 i_1}\cdots{}_T\pi_{\underline{s}_T i_T}\right)$$
$$= \delta\;_1\lambda(\underline{s}_1)\Gamma\;_2\lambda(\underline{s}_2)\cdots\Gamma\;_T\lambda(\underline{s}_T)\mathbf{1}',$$

where

$${}_t\lambda(\underline{s}) = \mathrm{diag}\left({}_t\pi_{\underline{s}1},\ldots,{}_t\pi_{\underline{s}m}\right).$$

In maximizing the likelihood in order to estimate parameters, one must observe all the constraints noted above.

3.3.2 Marginal properties and cross-correlations

Since $\{S_{tj} : t \in \mathbf{N}\}$ is, for each j, a binomial-hidden Markov model, the mean, variance, autocorrelation and distributional properties of $\{S_{tj} : t \in \mathbf{N}\}$ are exactly as derived in the preceding chapter. For instance, the mean and variance are given by

$$\mathrm{E}(S_{tj}) = n_t \sum_{i=1}^{m} \delta_i p_{ij} = n_t \delta p'_{(j)}$$

and

$$\mathrm{Var}(S_{tj})$$
$$= n_t(n_t-1)\sum_i \delta_i p_{ij}^2 + n_t\sum_i \delta_i p_{ij} - \left(n_t\sum_i \delta_i p_{ij}\right)^2$$
$$= n_t(n_t-1)\delta P_{(j)}p'_{(j)} + n_t\delta p'_{(j)} - n_t^2(\delta p'_{(j)})^2$$
$$= n_t^2\left(\delta P_{(j)}p'_{(j)} - (\delta p'_{(j)})^2\right) + n_t\left(\delta p'_{(j)} - \delta P_{(j)}p'_{(j)}\right), \quad (3.7)$$

where we define $p_{(j)} = (p_{1j}, p_{2j},\ldots,p_{mj})$ and $P_{(j)} = \mathrm{diag}(p_{(j)})$. In order to determine the cross-correlations $\mathrm{Corr}(S_{t1}, S_{t+k,2})$ we therefore need in addition only $\mathrm{E}(S_{t1}S_{t+k,2})$. (There is no loss of generality in considering only categories 1 and 2, as the categories can if necessary be renumbered.) We deal with the cases $k=0$ and $k>0$ separately.

Firstly,

$$\mathrm{E}(S_{t1}S_{t2}) = \sum_{i=1}^{m} \delta_i \mathrm{E}(S_{t1}S_{t2}\mid C_t=i) = \sum_i \delta_i n_t(n_t-1)p_{i1}p_{i2},$$

since the conditional joint distribution of S_{t1} and S_{t2} is multinomial

(more precisely, trinomial). Hence
$$E(S_{t1}S_{t2}) = n_t(n_t - 1)\delta P_{(1)}p'_{(2)}$$
and
$$\text{Cov}(S_{t1}, S_{t2}) = n_t(n_t - 1)\delta P_{(1)}p'_{(2)} - n_t^2(\delta p'_{(1)})(\delta p'_{(2)}),$$
which yields the correlation of S_{t1} and S_{t2} on division by the appropriate expressions for the standard deviations of S_{t1} and S_{t2}: see equation (3.7) above.

Secondly, note that for $k \in \mathbf{N}$

$$\begin{aligned}
E(S_{t1}S_{t+k,2}) &= \sum_{i,j=1}^{m} \delta_i \gamma_{ij}(k) E(S_{t1}S_{t+k,2} \mid C_t = i, C_{t+k} = j) \\
&= \sum_{i,j} \delta_i \gamma_{ij}(k)(n_t p_{i1})(n_{t+k} p_{j2}) \\
&= n_t n_{t+k} \delta P_{(1)} \Gamma^k p'_{(2)},
\end{aligned}$$

and
$$\text{Cov}(S_{t1}, S_{t+k,2}) = n_t n_{t+k} \left(\delta P_{(1)} \Gamma^k p'_{(2)} - (\delta p'_{(1)})(\delta p'_{(2)}) \right).$$

Division by the standard deviations of S_{t1} and $S_{t+k,2}$ then gives us the correlation of S_{t1} and $S_{t+k,2}$, which is:

$$(1 + \alpha_1/n_t)^{-1/2}(1 + \alpha_2/n_{t+k})^{-1/2}$$
$$\times \frac{\delta P_{(1)} \Gamma^k p'_{(2)} - (\delta p'_{(1)})(\delta p'_{(2)})}{\left(\delta P_{(1)} p'_{(1)} - (\delta p'_{(1)})^2 \right)^{1/2} \left(\delta P_{(2)} p'_{(2)} - (\delta p'_{(2)})^2 \right)^{1/2}},$$

where
$$\alpha_j = \frac{\delta p'_{(j)} - \delta P_{(j)} p'_{(j)}}{\delta P_{(j)} p'_{(j)} - (\delta p'_{(j)})^2}.$$

In the case $m = 2$, i.e. if the underlying Markov chain has two states, we have:

$$\begin{aligned}
\delta P_{(j)} p'_{(j)} - (\delta p'_{(j)})^2 &= \delta_1 \delta_2 (p_{2j} - p_{1j})^2 \\
\delta p'_{(j)} - \delta P_{(j)} p'_{(j)} &= \delta_1 p_{1j}(1 - p_{1j}) + \delta_2 p_{2j}(1 - p_{2j}) \\
\delta P_{(1)} p'_{(2)} - (\delta p'_{(1)})(\delta p'_{(2)}) &= \delta_1 \delta_2 (p_{21} - p_{11})(p_{22} - p_{12}) \\
\delta P_{(1)} p'_{(2)} &= \delta_1 p_{11} p_{12} + \delta_2 p_{21} p_{22} \\
\delta P_{(1)} \Gamma^k p'_{(2)} - (\delta p'_{(1)})(\delta p'_{(2)}) &= \delta_1 \delta_2 (p_{21} - p_{11})(p_{22} - p_{12}) w^k.
\end{aligned}$$

(As usual, w denotes $1 - \gamma_1 - \gamma_2$.) Hence we have for $m = 2$ the following results for the variances and covariances, and the cross-correlation of order k:

$\mathrm{Var}(S_{tj})$
$= n_t^2 \left(\delta P_{(j)} p'_{(j)} - (\delta p'_{(j)})^2\right) + n_t(\delta p'_{(j)} - \delta P_{(j)} p'_{(j)})$
$= n_t^2 \delta_1 \delta_2 (p_{2j} - p_{1j})^2 + n_t \{\delta_1 p_{1j}(1 - p_{1j}) + \delta_2 p_{2j}(1 - p_{2j})\};$

$\mathrm{Cov}(S_{t1}, S_{t2})$
$= n_t^2 \left(\delta P_{(1)} p'_{(2)} - \delta p'_{(1)} \delta p'_{(2)}\right) - n_t \delta P_{(1)} p'_{(2)}$
$= n_t^2 \delta_1 \delta_2 (p_{21} - p_{11})(p_{22} - p_{12}) - n_t (\delta_1 p_{11} p_{12} + \delta_2 p_{21} p_{22});$

and, for all $k \in \mathbf{N}$:

$\mathrm{Cov}(S_{t1}, S_{t+k,2}) = n_t n_{t+k} \left(\delta P_{(1)} \Gamma^k p'_{(2)} - \delta p'_{(1)} \delta p'_{(2)}\right)$
$= n_t n_{t+k} \delta_1 \delta_2 (p_{21} - p_{11})(p_{22} - p_{12}) w^k$

and

$\mathrm{Corr}(S_{t1}, S_{t+k,2})$
$= n_t n_{t+k} \delta_1 \delta_2 (p_{21} - p_{11})(p_{22} - p_{12}) w^k / (\mathrm{Var}(S_{t1}) \mathrm{Var}(S_{t+k,2}))^{1/2}$
$= (1 + \alpha_1/n_t)^{-1/2} (1 + \alpha_2/n_{t+k})^{-1/2}$
$\quad \times \mathrm{sgn}((p_{21} - p_{11})(p_{22} - p_{12})) w^k,$

where for $j = 1, 2$

$$\alpha_j = \frac{\delta_1 p_{1j}(1 - p_{1j}) + \delta_2 p_{2j}(1 - p_{2j})}{\delta_1 \delta_2 (p_{2j} - p_{1j})^2}.$$

From the above it follows that, if n_t is constant with respect to t (and m is 2), the cross-correlation of order $k \in \mathbf{N}$ depends on k only through w^k, and therefore falls off geometrically with increasing k.

3.3.3 A model for categorical time series

We now consider the case $n_t = 1$ (and m general), the model for categorical time series. This case may be of considerable practical interest. Apart from Markov chains of first order or higher, the only other approaches that seem to be available for general categorical series are the observation-driven models of Kaufmann (1987) and Fahrmeir and Kaufmann (1987) (see section 1.7), and the 'sequency domain' approach of Stoffer, which was described in section 1.10.

The models we discuss here, being parameter-driven, may be applicable to data for which observation-driven models are inappropriate.

The general results of section 3.3.2, for means, variances and covariances, specialize as follows to the case $n_t = 1$:

$$E(S_{tj}) = \delta p'_{(j)}$$
$$\text{Var}(S_{tj}) = \delta p'_{(j)} - (\delta p'_{(j)})^2$$
$$\text{Cov}(S_{t1}, S_{t2}) = -(\delta p'_{(1)})(\delta p'_{(2)}),$$

and, for all $k \in \mathbf{N}$,

$$\text{Cov}(S_{t1}, S_{t+k,2}) = \delta P_{(1)} \Gamma^k p'_{(2)} - (\delta p'_{(1)})(\delta p'_{(2)}).$$

The corresponding expressions for the cross-correlations follow in obvious fashion:

$$\text{Corr}(S_{t1}, S_{t2}) = \frac{-(\delta p'_{(1)})(\delta p'_{(2)})}{\left(\delta p'_{(1)} - (\delta p'_{(1)})^2\right)^{1/2} \left(\delta p'_{(2)} - (\delta p'_{(2)})^2\right)^{1/2}}$$

and, for $k \in \mathbf{N}$,

$$\text{Corr}(S_{t1}, S_{t+k,2}) = \frac{\delta P_{(1)} \Gamma^k p'_{(2)} - (\delta p'_{(1)})(\delta p'_{(2)})}{\left(\delta p'_{(1)} - (\delta p'_{(1)})^2\right)^{1/2} \left(\delta p'_{(2)} - (\delta p'_{(2)})^2\right)^{1/2}}.$$

It is convenient to repeat here the definitions of some of the notation being used throughout section 3.3, in order that we may note how it specializes in the present case. Given $C_t = i$, the probability that S_{tj}, the jth component of $\underline{S}_t$, equals 1 (and $S_{tl} = 0$ for all $l \neq j$) is p_{ij}. Accordingly $\sum_{j=1}^{q} p_{ij} = 1$ for each i from 1 to m. The vector $p_{(j)}$ is defined as $(p_{1j}, \ldots, p_{mj})$, and the matrix $P_{(j)}$ as $\text{diag}(p_{(j)})$. Because $\sum_{j=1}^{q} p_{ij} = 1$, we have $\sum_{j=1}^{q} p_{(j)} = \mathbf{1}$ and $\sum_{j=1}^{q} P_{(j)} = I_m$.

Here the state-dependent probabilities $_t\pi_{\underline{s}i}$ and the matrix expression for the likelihood simplify considerably. If the jth component of $\underline{s}$ is 1 (and the others are therefore 0) we have

$$_t\pi_{\underline{s}i} = p_{ij},$$

and the subscript t is clearly unnecessary. It follows that

$$\lambda(\underline{s}) = \text{diag}(\pi_{\underline{s}1}, \ldots, \pi_{\underline{s}m})$$
$$= \text{diag}(p_{1j}, \ldots, p_{mj})$$
$$= P_{(j)},$$

where again $\underline{s}$ is the vector with jth component 1 and the others 0. Hence the likelihood of observing categories $j_1, j_2, \ldots, j_T$ at times $1, 2, \ldots, T$ is given by

$$\delta P_{(j_1)} \Gamma P_{(j_2)} \Gamma \cdots P_{(j_T)} \mathbf{1}'.$$

This implies, for instance, that the probability of observing category j at time t, given category l is observed at time $t-1$, is

$$\frac{\delta P_{(l)} \Gamma P_{(j)} \mathbf{1}'}{\delta P_{(l)} \mathbf{1}'}. \quad (3.8)$$

The models of this section will be used in section 4.6 to model a series of wind directions in the form of the conventional 16 categories, i.e. the 16 points of the compass.

3.4 Multivariate models

Consider now the following multivariate extension of hidden Markov models.

Let $\{C_t\}$ be the usual first-order Markov chain on m states, and suppose that, conditional on $C^{(T)}$, the Tq random variables $\{S_{tj} : t=1, \ldots, T, \; j=1, \ldots, q\}$ are mutually independent. That is, we consider the q time series $\{S_{tj} : t=1, \ldots, T\}$, and assume that there is contemporaneous conditional independence as well as the usual conditional independence along time. An example of such a model is the multisite precipitation model of Zucchini and Guttorp (1991): in the application they describe, five binary time series represent the presence or absence of rain at each of five sites linked by a common 'climate process' $\{C_t\}$. In section 3.4 we shall discuss *inter alia* the properties of models slightly more general than theirs, involving binomial distributions rather than merely Bernoulli, and those of similar models involving Poisson distributions.

A more general model yet could be obtained by relaxing the assumption of contemporaneous conditional independence, and in fact such a modification is suggested by Zucchini and Guttorp in the context of rainfall sites situated in close proximity: in that case the assumption of conditional independence across sites may be unrealistic. The multinomial-hidden Markov model of section 3.3 is an example of a multivariate model in which contemporaneous conditional independence is not assumed. A further example of this type would be a model in which the conditional distribution of the random vector $\underline{S}_t = (S_{t1}, S_{t2}, \ldots, S_{tq})$ is a multivariate Poisson with parameters determined by the current state of the underlying

Markov chain. Such a model is described in section 3.4.3.

3.4.1 The likelihood function for multivariate models

We begin by giving the matrix expression for the likelihood of observations $\underline{s}_1, \ldots, \underline{s}_T$ from a general multivariate hidden Markov model, i.e. one in which contemporaneous conditional independence is not assumed. This has effectively already been established in section 3.3. With the definitions

$$_t\pi_{\underline{s}i} = P(\underline{S}_t = \underline{s} \mid C_t = i)$$

and

$$_t\lambda(\underline{s}) = \text{diag}(_t\pi_{\underline{s}1}, \ldots, _t\pi_{\underline{s}m}),$$

we have as the likelihood

$$\delta_1 \lambda(\underline{s}_1) \Gamma_2 \lambda(\underline{s}_2) \cdots \Gamma_T \lambda(\underline{s}_T) \mathbf{1}'.$$

In the case of contemporaneous conditional independence, the state-dependent probabilities $_t\pi_{\underline{s}i}$ are given by a product:

$$_t\pi_{\underline{s}_t i} = \prod_{j=1}^{q} P(S_{tj} = s_{tj} \mid C_t = i).$$

The multisite precipitation model of Zucchini and Guttorp is a case in point. There the random variables S_{tj} are binary, and if p_{ij} denotes

$$P(S_{tj} = 1 \mid C_t = i) = 1 - P(S_{tj} = 0 \mid C_t = i),$$

then

$$_t\pi_{\underline{s}_t i} = \prod_{j=1}^{q} p_{ij}^{s_{tj}} (1 - p_{ij})^{1-s_{tj}}.$$

3.4.2 Cross-correlations of models assuming contemporaneous conditional independence

For each j, $\{S_{tj} : t \in \mathbf{N}\}$ is a univariate hidden Markov model. We shall therefore say little about its marginal properties, i.e. the mean, variance, autocorrelation and distributional properties of $\{S_{tj} : t \in \mathbf{N}\}$ for a specific j. We consider here the cross-correlation structure of models with contemporaneous conditional independence (as well as conditional independence along time) and either a Poisson or a binomial conditional distribution for each S_{tj}.

In the Poisson case, let S_{tj} have mean λ_{ij} if $C_t=i$. Define $\lambda_{(j)} = (\lambda_{1j}, \lambda_{2j}, \ldots, \lambda_{mj})$ and $\Lambda_{(j)} = \text{diag}(\lambda_{(j)})$. We then have for all $k \in \mathbf{N}$:

$$\begin{aligned}
\mathrm{E}(S_{t1}S_{t+k,2}) &= \sum_{i,j=1}^{m} \delta_i \gamma_{ij}(k) \mathrm{E}(S_{t1}S_{t+k,2} \mid C_t=i, C_{t+k}=j) \\
&= \sum_{i,j} \delta_i \lambda_{i1} \gamma_{ij}(k) \lambda_{j2} \\
&= \delta \Lambda_{(1)} \Gamma^k \lambda'_{(2)}; \quad (3.9) \\
\mathrm{Cov}(S_{t1}, S_{t+k,2}) &= \delta \Lambda_{(1)} \Gamma^k \lambda'_{(2)} - (\delta \lambda'_{(1)})(\delta \lambda'_{(2)}).
\end{aligned}$$

Because of the assumed contemporaneous independence the conclusions drawn above are valid also for $k=0$:

$$\mathrm{E}(S_{t1}S_{t2}) = \sum_{i=1}^{m} \delta_i \, \mathrm{E}(S_{t1}S_{t2} \mid C_t=i) = \sum_{i=1}^{m} \delta_i \lambda_{i1} \lambda_{i2} = \delta \Lambda_{(1)} \lambda'_{(2)},$$

and

$$\mathrm{Cov}(S_{t1}, S_{t2}) = \delta \Lambda_{(1)} \lambda'_{(2)} - (\delta \lambda'_{(1)})(\delta \lambda'_{(2)}).$$

Using the expression for the variance derived in section 2.4.1, that is,

$$\mathrm{Var}(S_{tj}) = \lambda_{(j)} D \lambda'_{(j)} + \delta \lambda'_{(j)} - (\delta \lambda'_{(j)})^2 = \delta \Lambda_{(j)} \lambda'_{(j)} + \delta \lambda'_{(j)} - (\delta \lambda'_{(j)})^2,$$

we can therefore write down the correlation of S_{t1} and $S_{t+k,2}$ for all nonnegative integers k.

The special case $m=2$ of this model (i.e. of the Poisson-based model with contemporaneous independence) yields the following result for the cross-covariances, valid for all nonnegative integers k:

$$\mathrm{Cov}(S_{t1}, S_{t+k,2}) = \delta_1 \delta_2 (\lambda_{21} - \lambda_{11})(\lambda_{22} - \lambda_{12}) w^k,$$

where $w = 1 - \gamma_1 - \gamma_2$. From this and the corresponding result for the variances (see section 2.4.1),

$$\mathrm{Var}(S_{tj}) = \delta_1 \delta_2 (\lambda_{2j} - \lambda_{1j})^2 + \delta \lambda'_{(j)},$$

we can write down an expression for the cross-correlation $\mathrm{Corr}(S_{t1}, S_{t+k,2})$, valid for all nonnegative integers k.

The binomial version of the above model requires that the conditional distribution of S_{tj} ($t=1,\ldots,T$; $j=1,\ldots,q$) be binomial with parameters n_{tj} (known) and p_{ij} if $C_t=i$. The results are very similar to the Poisson case. For nonnegative integers k

$$\mathrm{Cov}(S_{t1}, S_{t+k,2}) = n_{t1} n_{t+k,2} \left(\delta P_{(1)} \Gamma^k p'_{(2)} - (\delta p'_{(1)})(\delta p'_{(2)}) \right).$$

Since we have from section 2.4.2 that

$$\text{Var}(S_{tj}) = n_{tj}^2 \left(\delta P_{(j)} p'_{(j)} - (\delta p'_{(j)})^2 \right) + n_{tj} \left(\delta p'_{(j)} - \delta P_{(j)} p'_{(j)} \right),$$

the cross-correlation is available in general. Two special cases of such binomial-based models may be of interest: $m = 2$, and $n_{tj} = 1$ for all t and j. (The latter case corresponds to the multisite precipitation model.)

For $m = 2$,

$$\begin{aligned}
\text{Cov}(S_{t1}, S_{t+k,2}) &= n_{t1} n_{t+k,2}\, \delta P_{(1)} \begin{pmatrix} \delta_2 & -\delta_2 \\ -\delta_1 & \delta_1 \end{pmatrix} p'_{(2)} w^k \\
&= n_{t1} n_{t+k,2}\, \delta_1 \delta_2 (p_{21} - p_{11})(p_{22} - p_{12}) w^k,
\end{aligned}$$

which is essentially the same conclusion as holds only for *positive* integers k in the case of multinomial-hidden Markov models. With the corresponding result for the variances (see section 2.4.2),

$$\text{Var}(S_{tj})$$
$$= n_{tj}^2 \delta_1 \delta_2 (p_{2j} - p_{1j})^2 + n_{tj}\left(\delta_1 p_{1j}(1 - p_{1j}) + \delta_2 p_{2j}(1 - p_{2j}) \right),$$

this yields the cross-correlations $\text{Corr}(S_{t1}, S_{t+k,2})$ for nonnegative integer k.

If $n_{tj} = 1$ for all t and j (and m is general), then we have the results:

$$\begin{aligned}
\text{Cov}(S_{t1}, S_{t+k,2}) &= \delta P_{(1)} \Gamma^k p'_{(2)} - (\delta p'_{(1)})(\delta p'_{(2)}) \\
\text{Var}(S_{tj}) &= \delta p'_{(j)} - (\delta p'_{(j)})^2 \\
\text{Corr}(S_{t1}, S_{t+k,2}) &= \frac{\delta P_{(1)} \Gamma^k p'_{(2)} - (\delta p'_{(1)})(\delta p'_{(2)})}{\left(\delta p'_{(1)} - (\delta p'_{(1)})^2 \right)^{1/2} \left(\delta p'_{(2)} - (\delta p'_{(2)})^2 \right)^{1/2}}.
\end{aligned}$$

3.4.3 Cross-correlations of models not assuming contemporaneous conditional independence

We have already discussed, in section 3.3, the cross-correlations of one class of models which does not assume contemporaneous conditional independence, the multinomial-hidden Markov models. We now indicate, as a further example, how the cross-correlation function can be obtained for a bivariate hidden Markov model in which the conditional distribution is a bivariate Poisson. (Because of the independence along time, the only cross-correlation that differs from those obtained for the Poisson-based models discussed in section 3.4.2 is the cross-correlation at lag zero.) Suppose then

MULTIVARIATE MODELS   125

that, if $C_t = i$, S_{t1} and S_{t2} have the bivariate distribution with joint generating function

$$\exp(\lambda_{i1}(u-1) + \lambda_{i2}(v-1) + a_i(u-1)(v-1)),$$

where λ_{i1}, λ_{i2} and a_i are all positive. This bivariate generating function is of the form implied by the multivariate Poisson distribution of Teicher (1954), and in turn implies marginal means λ_{i1} and λ_{i2}, and covariance a_i. (Clearly we must require that $a_i \leq (\lambda_{i1}\lambda_{i2})^{1/2}$.) Then for all $k \in \mathbf{N}$ we have, as in equation (3.9), that

$$E(S_{t1}S_{t+k,2}) = \sum_{i,j=1}^{m} \delta_i \gamma_{ij}(k) \lambda_{i1} \lambda_{j2} = \delta \Lambda_{(1)} \Gamma^k \lambda'_{(2)}.$$

The corresponding result for lag zero is

$$E(S_{t1}S_{t2}) = \sum_{i=1}^{m} \delta_i (a_i + \lambda_{i1}\lambda_{i2}).$$

Since expressions are available from section 2.4.1 for the mean and variance of S_{tj}, the cross-covariances and -correlations can therefore be computed.

3.4.4 Multivariate models with time lags

Suppose we consider models in which there is the usual Markov chain $\{C_t\}$, and a vector of observations of which some depend on C_{t-1} and some on C_t. To illustrate the nature of such models we consider in particular one in which there are only two observations at each time point. We assume that, conditional on $C_0, C_1, \ldots, C_T$, the random variables $S_1, \ldots, S_T$ and $U_1, \ldots, U_T$ are mutually independent and the (conditional) distributions of S_t and U_t are given by:

$$\begin{aligned} P(S_t = s \mid C_{t-1} = i) &= \pi_{si} \\ P(U_t = u \mid C_t = j) &= \sigma_{uj}. \end{aligned} \quad (3.10)$$

The likelihood of T consecutive observations $(S_1, U_1), \ldots, (S_T, U_T)$ is then seen to be

$$\sum_{i_0, i_1, \ldots, i_T = 1}^{m} \left(\delta_{i_0} \gamma_{i_0 i_1} \cdots \gamma_{i_{T-1} i_T} \right) \left(\pi_{s_1 i_0} \cdots \pi_{s_T i_{T-1}} \right) \left(\sigma_{u_1 i_1} \cdots \sigma_{u_T i_T} \right)$$

$$= \delta A_1 A_2 \cdots A_T \mathbf{1}',$$

where A_t is defined, for $i, j = 1, 2, \ldots, m$, by
$$(A_t)_{ij} = \gamma_{ij} \pi_{s_t;i} \sigma_{u_t;j}.$$

(In fact the form of A_t suggests that it is not difficult to generalize this result to a model in which the distributions of S_t and U_t depend on both C_{t-1} and C_t. If we define:
$$\pi_{s;ij} = \mathrm{P}(S_t = s \mid C_{t-1} = i, C_t = j)$$
$$\sigma_{u;ij} = \mathrm{P}(U_t = u \mid C_{t-1} = i, C_t = j)$$
$$(B_t)_{ij} = \gamma_{ij} \pi_{s_t;ij} \sigma_{u_t;ij},$$

then the likelihood can be written as $\delta B_1 B_2 \cdots B_T \mathbf{1}'$. We shall, however, not pursue this generalization, as a more efficient way of obtaining properties of such a process would to be redefine the parameter process to be $\{X_t\}$, where $X_t = (C_{t-1}, C_t)$, and treat the model as a standard hidden Markov model of the type described in Chapter 2.)

For the models defined by equations (3.10), there is nothing new that needs to be said about the marginal properties of $\{U_t\}$, or for that matter $\{S_t\}$, although in the latter case it is notationally convenient to define $R_{t-1} = S_t$ and then consider the standard hidden Markov model $\{R_t\}$. This redefinition is also useful for computing the cross-correlations of $\{S_t\}$ and $\{U_t\}$, because one can then apply the results of section 3.4.2 to $\{R_t\}$ and $\{U_t\}$.

3.4.5 Multivariate models in which some variables are discrete and others continuous

Some multivariate time series have both discrete- and continuous-valued components, e.g. the bivariate series comprising a binary series of wet days and dry days at a particular site and a continuous-valued series of measurements on some other climatic variable, such as relative humidity, solar radiation or maximum temperature. Another example of such a bivariate series would be wind direction (in categories) and wind speed.

Such 'mixed' bivariate series can also be modelled by means of a hidden Markov model by conditioning their distributions on a single Markov chain, $\{C_t\}$. We assume that there is the usual conditional independence along time, but not necessarily contemporaneously: that is, we do not in general assume that, given the current state of the Markov chain, the discrete variable and the continuous one are independent. Conditional on $\{C_t\}$, let the joint distribution of S_t (discrete) and U_t (continuous) be represented

MULTIVARIATE MODELS 127

by the joint probability mass/density function $f_i(s,u)$. If there is indeed contemporaneous conditional independence, then

$$f_i(s,u) = \pi_{si} g_i(u), \qquad (3.11)$$

where $\pi_{\cdot i}$ and g_i are (in general) the marginal probability mass and density functions deducible from f_i. What follows, however, refers to the more general model unless otherwise indicated.

The likelihood is given by:

$$\sum_{i_1,\ldots,i_T=1}^{m} \left(\delta_{i_1}\gamma_{i_1 i_2}\cdots\gamma_{i_{T-1} i_T}\right) \prod_{t=1}^{T} f_{i_t}(s_t, u_t)$$
$$= \delta\lambda(s_1, u_1)\Gamma\lambda(s_2, u_2)\cdots\Gamma\lambda(s_T, u_T)\mathbf{1}',$$

where $\lambda(s,u) = \text{diag}(f_1(s,u),\ldots,f_m(s,u))$. The marginal properties of $\{S_t\}$, the discrete component, involve nothing essentially new. The marginal properties of $\{U_t\}$ merely require the use of integration rather than summation in various places. For instance

$$\begin{aligned}\text{E}(U_t^k) &= \text{E}\left(\text{E}(U_t^k \mid C^{(T)})\right) \\ &= \sum_{i=1}^{m}\delta_i \int_{-\infty}^{\infty} u^k g_i(u)\,\mathrm{d}u.\end{aligned}$$

The cross-correlations can therefore be computed once we have found $\text{E}(S_t U_{t+k})$ and $\text{E}(U_t S_{t+k})$ for nonnegative integers k. For $k \in \mathbf{N}$ we have:

$$\begin{aligned}\text{E}(S_t U_{t+k}) &= \sum_{i,j=1}^{m} \delta_i \gamma_{ij}(k) \text{E}(S_t \mid C_t=i)\,\text{E}(U_{t+k} \mid C_{t+k}=j) \\ &= \sum_{i,j} \delta_i \gamma_{ij}(k)\lambda_i \mu_j \\ &= \delta \Lambda \Gamma^k \mu',\end{aligned}$$

where λ_i and μ_i are the conditional means of S_t and U_t given $C_t = i$, and $\lambda = (\lambda_1,\ldots,\lambda_m)$, $\mu = (\mu_1,\ldots,\mu_m)$ and $\Lambda = \text{diag}(\lambda)$. Similarly, for $k \in \mathbf{N}$, we have

$$\text{E}(U_t S_{t+k}) = \delta M \Gamma^k \lambda',$$

where M denotes $\text{diag}(\mu)$. The cross-correlation at lag zero requires

$$\text{E}(S_t U_t) = \sum_{i=1}^{m} \delta_i \text{E}(S_t U_t \mid C_t=i),$$

with this last conditional expectation being given by

$$\sum_s \int_{-\infty}^{\infty} su f_i(s,u)\,\mathrm{d}u,$$

which equals $\lambda_i \mu_i$ if assumption (3.11) holds.

This class of models can fairly routinely be extended to cater for several discrete components and several continuous. It is also possible to allow the joint distribution of S_t and U_t to depend on t as well as on the state currently occupied by the Markov chain.

3.5 Models with state-dependent probabilities depending on covariates

Hidden Markov models can be modified to allow for the influence of covariates by postulating dependence of the state-dependent probabilities $_t\pi_{si}$ on those covariates. This opens the way for such models to incorporate time trend and seasonality, for instance. We take $\{C_t\}$ to be the usual Markov chain, and we suppose, in the case of Poisson-hidden Markov models, that the conditional mean $_t\lambda_i$ depends on the (row) vector x_t of q covariates, for instance as follows:

$$\log\,{}_t\lambda_i = \beta_i x_t'.$$

(We continue here to use the convention that the subscript t before a symbol indicates time-dependence of the quantity concerned.) In the case of binomial-hidden Markov models, the corresponding assumption is that

$$\operatorname{logit}\,{}_t p_i = \beta_i x_t'.$$

The elements of x_t could include a constant, time (t), sinusoidal components expressing seasonality (for example $\cos(2\pi t/r)$ and $\sin(2\pi t/r)$ for some positive integer r), and any other relevant covariates. A binomial-hidden Markov model with

$$\operatorname{logit}\,{}_t p_i = \beta_{i1} + \beta_{i2} t + \beta_{i3}\cos(2\pi t/r) + \beta_{i4}\sin(2\pi t/r) + \beta_{i5} y_t + \beta_{i6} z_t$$

allows for a (logistic-linear) time trend, r-period seasonality and the influence of covariates y_t and z_t, in the state-dependent 'success probabilities' $_t p_i$. Additional sine–cosine pairs can if necessary be included, to model more complex seasonal patterns. Similar models for the log of the conditional mean $_t\lambda_i$ are possible in the Poisson-hidden Markov case. Clearly link functions other than the canonical ones used here could be considered too.

The expression $\delta_1\lambda(s_1)\Gamma_2\lambda(s_2)\cdots\Gamma_T\lambda(s_T)\mathbf{1}'$ for the likelihood of T consecutive observations $s_1,\ldots,s_T$ remains valid for these models involving covariates: what changes is the precise definition of ${}_t\pi_{si}$, and hence of

$$_t\lambda(s) = \text{diag}({}_t\pi_{s1},\ldots,{}_t\pi_{sm}).$$

Expressions for moments, including autocorrelations, are found by the usual methods, although here the autocorrelations are less useful than they are in the case of stationary models. For the Poisson-hidden Markov models with covariates we have

$$\begin{aligned}
\text{E}(S_t) &= \sum_i \delta_i \, {}_t\lambda_i \\
\text{E}(S_t^2) &= \sum_i \delta_i \left({}_t\lambda_i + ({}_t\lambda_i)^2\right) \\
\text{E}(S_t S_{t+k}) &= \sum_{i,j} \delta_i \gamma_{ij}(k) \, {}_t\lambda_i \, {}_{t+k}\lambda_j ,
\end{aligned}$$

where $k \in \mathbf{N}$ and in all cases ${}_t\lambda_i = \exp(\beta_i x_t')$. For the binomial-hidden Markov model with covariates:

$$\begin{aligned}
\text{E}(S_t) &= n_t \sum_i \delta_i \, {}_t p_i \\
\text{E}(S_t^2) &= \sum_i \delta_i \left(n_t \, {}_t p_i (1 - {}_t p_i) + n_t^2 \, {}_t p_i^2\right) \\
\text{E}(S_t S_{t+k}) &= n_t n_{t+k} \sum_{i,j} \delta_i \gamma_{ij}(k) \, {}_t p_i \, {}_{t+k} p_j ,
\end{aligned}$$

where $k \in \mathbf{N}$ and ${}_t p_i = \exp(\beta_i x_t') / (1 + \exp(\beta_i x_t'))$. Expressions for $\text{Var}(S_t)$, $\text{Cov}(S_t, S_{t+k})$ and $\text{Corr}(S_t, S_{t+k})$ then follow for the two kinds of model.

3.6 Models in which the Markov chain is homogeneous but not assumed stationary

If the Markov chain $\{C_t\}$ underlying a hidden Markov model is homogeneous but not necessarily stationary, the model is of the type discussed by Leroux and Puterman (1992) and described in section 2.7. It will be recalled from that section that the likelihood can in that case be maximized by taking the Markov chain to start (with probability 1) at a particular state, and applying the EM algorithm to estimate the transition probabilities γ_{ij} and the remaining parameters, which in the application of Leroux and Puterman are the means λ_j of the Poisson conditional distributions. For practical purposes (such as forecasting), except perhaps in the case of short series exhibiting strong serial dependence, models of this kind differ only slightly from those in which the Markov chain is assumed to be stationary. Since such nonstationary models have been discussed in some detail by the authors mentioned, we shall not dwell on the topic here.

3.7 Models in which the Markov chain is nonhomogeneous

A further way in which one can accommodate time trend and seasonality in hidden Markov models is to drop the assumption that the Markov chain is homogeneous, and assume instead that the transition probabilities are a function of time. More generally, the transition probabilities can be modelled as depending on one or more covariates, not necessarily time but any variables that are considered relevant.

The incorporation of covariates into the Markov chain is not as straightforward as incorporating them into the state-dependent probabilities. One reason why it could be worthwhile, however, is that the resulting Markov chain may have a useful substantive interpretation: e.g. as a 'weather process' which is itself complex but determines rainfall probabilities at several sites in fairly simple fashion.

Hughes (1993) describes nonhomogeneous hidden Markov models for relating synoptic atmospheric patterns to local hydrologic phenomena. His models are based on the following two assumptions:

$$P\left(S_t \mid C^{(T)}, S^{(t-1)}, X^{(T)}\right) = P(S_t \mid C_t),$$

and

$$P\left(C_t \mid C^{(t-1)}, X^{(T)}\right) = P(C_t \mid C_{t-1}, X_t).$$

In his application S_t is a vector of measurements of the local process at time t (presence or absence of precipitation at a number of sites), C_t is the underlying unobserved Markov chain (the 'weather state' at time t), and X_t is a covariate (the 'measurement — or summary — of atmospheric data' at time t). Here $S^{(T)}$ is the set of all observations S_t ($t=1, 2, \ldots, T$), i.e. the history from time 1 to T, $C^{(t-1)}$ is the history of the (unobserved) states of the underlying Markov chain from 1 to $t-1$, and $X^{(T)}$ the history of the covariates X_t from 1 to T.

The form of the dependence of the transition probabilities on the covariates has to be specified in a way that is appropriate to the particular application. Hughes therefore discusses several different parametrizations for the transition probabilities of the weather states C_t, and for the conditional distribution of the precipitation S_t, including, in the latter case, a model that incorporates spatial dependence in the occurrence of precipitation at different rain sites.

In most cases parameter estimation is carried out by numerical maximization of the likelihood, which is of the form:

$$\delta(x_1)B(s_1)A(x_2)B(s_2)\cdots A(x_T)B(s_T)\mathbf{1}'$$

where:

$x_1, x_2, \ldots, x_T$ are the observed values of the covariate;

$s_1, s_2, \ldots, s_T$ are the observed values of the process $\{S_t\}$;

$A(x)$ is the $m \times m$ matrix with entries $A_{ij}(x) = \mathrm{P}(C_t = j \mid C_{t-1} = i, X_t = x)$;

$B(s)$ is the $m \times m$ diagonal matrix with ith diagonal entry $B_{ii}(s) = \mathrm{P}(S_t = s \mid C_t = i)$; and

$\delta(x)$ is the stationary distribution implied by the transition probability matrix $A(x)$.

Although Hughes (1993) concentrates on the problem of modelling precipitation, the general nonhomogeneous hidden Markov structure that he describes, and the techniques he develops, are potentially applicable to a considerably wider class of problems.

We return now to the issue of modelling time trend and seasonality in hidden Markov models, and indicate here one possible way to modify the transition probabilities of the Markov chain to achieve this.

Consider a model based on a two-state Markov chain $\{C_t\}$ with

$$\mathrm{P}(C_t = 2 \mid C_{t-1} = 1) = {}_t\gamma_1,$$
$$\mathrm{P}(C_t = 1 \mid C_{t-1} = 2) = {}_t\gamma_2$$

and, for $i = 1, 2$

$$\mathrm{logit}\ {}_t\gamma_i = \beta_{(i)} x_t'.$$

For example, a model incorporating r-period seasonality is that with

$$\mathrm{logit}\ {}_t\gamma_i = \beta_{(i)1} + \beta_{(i)2} \cos(2\pi t/r) + \beta_{(i)3} \sin(2\pi t/r).$$

In general the above assumption on logit ${}_t\gamma_i$ implies that the transition probability matrix, for transitions between times $t-1$ and t, is given by

$$_t\Gamma = \begin{pmatrix} \frac{1}{1+\exp(\beta_{(1)}x'_t)} & \frac{\exp(\beta_{(1)}x'_t)}{1+\exp(\beta_{(1)}x'_t)} \\ \frac{\exp(\beta_{(2)}x'_t)}{1+\exp(\beta_{(2)}x'_t)} & \frac{1}{1+\exp(\beta_{(2)}x'_t)} \end{pmatrix}.$$

Extension of this model to the case $m > 2$ presents difficulties, but they appear not to be insuperable. Aitchison (1986, Chapter 6) presents several one-to-one transformations (e.g. the generalized, or 'additive', logistic transform) from $\mathbf{R}^n$ to the n-dimensional unit simplex. By applying such a transform to $n = m - 1$ appropriate linear functions $\beta x'_t$ of the covariates we can model the $m - 1$ off-diagonal transition probabilities in each row in a fashion consistent with the row sum constraint. Since this has to be done separately for each of the m rows of the transition probability matrix, the number of parameters used may be large unless some restrictions are imposed on the coefficients appearing in the linear functions.

One important difference between the class of models proposed here and other hidden Markov models (and a consequence of the nonhomogeneity) is that we cannot always assume there is a stationary distribution for the Markov chain. This problem arises when one or more of the covariates are functions of time, as in models for trend or seasonality. If necessary we therefore assume instead that there is some initial distribution δ at time $t=1$.

This has implications for the way in which parameters may be estimated. We now have to estimate three sets of parameters: the initial probabilities δ, the parameters appearing in the transition probabilities, and the parameters determining the state-dependent probabilities $_t\pi_{si}$. The unconditional likelihood is a convex linear combination of likelihood values conditioned on a particular initial state, and may for instance be maximized (as in the work of Leroux and Puterman) by choosing as the initial state that one which produces the largest maximized conditional likelihood. The EM algorithm may be used to estimate the other parameters. At the M-step the two parts of the complete-data log-likelihood can be maximized separately. The part involving the state-dependent probabilities may be maximized exactly as in the work of Leroux and Puterman, hence by closed-form expressions in the case of Poisson or binomial conditional distributions. The maximization of the part involving the transition probabilities is more complicated, and will need numerical solution except possibly in very special cases.

Alternatively, one can take an entirely numerical approach to

NUMBERS OF TRIALS AND NUMBERS OF SUCCESSES   133

the maximization of the likelihood conditional on the initial state of the Markov chain. This is done in the case of one of the models for the hourly wind direction series discussed in section 4.6, a hidden Markov model incorporating daily and annual cycles in the underlying two-state Markov chain.

3.8 Joint models for the numbers of trials and the numbers of successes in those trials

So far, when discussing binomial-hidden Markov models, we have always taken n_t, the number of trials at time t, to be a known constant. However, n_t could itself be an observation at time t on some nonnegative integer-valued random process $\{N_t\}$, for example a Poisson-hidden Markov model. One could then take $\{S_t\}$ to be such that, conditional on $\{N_t\}$, the process $\{S_t\}$ is a binomial-hidden Markov model with the numbers of trials n_t supplied by $\{N_t\}$, and driven by the same Markov chain as drives $\{N_t\}$ or even by another independent one. This includes as a special case the possibility that, given $\{N_t\}$, the observations S_t are independent binomial random variables.

To motivate this class of models, we consider an example thereof. Suppose that $\{N_t\}$ is a Poisson-hidden Markov model based on the two-state Markov chain $\{C_t\}$, with

$$\log{}_t\lambda_i = a_i + bt + c\cos(2\pi t/12) + d\sin(2\pi t/12),$$

for $i = 1, 2$ and $t = 1, \ldots, T$. Suppose further that, given $N^{(T)}$, the random variables S_t ($t = 1, \ldots, T$) are independent binomial $(N_t, {}_tp)$, with

$$\operatorname{logit}{}_tp = \alpha + \beta t. \tag{3.12}$$

The observations N_t and S_t could represent respectively the total number of births and the number of deliveries by a particular method, at a hospital in month t. Such a model for $\{N_t\}$ and $\{S_t\}$ would have a total of nine parameters, and would allow for time trend both in the number of births and in the proportion by the particular method of interest, for an annual cycle in the number of births, and for two underlying states which influence the mean number of births occurring at the hospital in a month. (Transport difficulties caused by weather or other factors could sometimes influence the number of births occurring at the hospital, and would be allowed for by the two states.) The two parameters α and β can be estimated, independently of the other parameters, by ordinary logistic regression, and the remaining seven parameters by numerical maximization of the Poisson-hidden Markov likelihood derived from the observations $N_1, \ldots, N_T$.

A modification of the above model which could perhaps prove useful in the application described is to add to the expression for logit $_tp$ in equation (3.12) a term involving n_t or some function thereof. This could be used to accommodate a 'busy period' effect: certain methods of delivery might be preferred by the obstetricians during particularly busy months. Other variations can similarly be incorporated into either the model $\{N_t\}$ or the model $\{S_t\}$.

A fairly general model of this kind is the following. The process $\{N_t\}$ is a Poisson-hidden Markov model with underlying stationary Markov chain $\{C_t\}$. Given $C^{(T)}$ and $N^{(T)}$, the random variables $S_1, \ldots, S_T$ are independent binomial with parameters N_t and $_tp_i$, where $C_t = i$. (The same Markov chain is taken to drive the two hidden Markov models involved.) The likelihood of the observations $N_1, \ldots, N_T$ and $S_1, \ldots, S_T$ is given by

$$\sum_{i_1,\ldots,i_T=1}^{m} \left(\delta_{i_1} \gamma_{i_1 i_2} \cdots \gamma_{i_{T-1} i_T}\right)$$
$$\times \left(_1\pi_{n_1 i_1} \cdots {_T}\pi_{n_T i_T}\right)\left(_1\nu_{s_1;n_1 i_1} \cdots {_T}\nu_{s_T;n_T i_T}\right),$$

where we define

$$_t\pi_{ni} = P(N_t = n \mid C_t = i) \quad \text{(the appropriate Poisson probability)}$$

and

$$_t\nu_{s;ni} = P(S_t = s \mid N_t = n, C_t = i) = \binom{n}{s} {_tp_i^s}(1 - {_tp_i})^{n-s}.$$

If we make the further definition that D_t is the diagonal matrix with ith diagonal element $_t\pi_{n_t i}\, _t\nu_{s_t;n_t i}$, we can therefore write the likelihood as

$$\delta D_1 \Gamma D_2 \cdots \Gamma D_T \mathbf{1}',$$

and use this expression for parameter estimation.

Another interesting class of models involves the use of two independent models, one for $\{N_t\}$ and one for $\{S_t\}$ given $\{N_t\}$. For example, one could use two independent hidden Markov models, or else a Markov regression model for $\{N_t\}$ and an independent binomial-hidden Markov model for $\{S_t\}$ given $\{N_t\}$. The assumption of independence allows one to estimate the parameters of the two models separately.

Such joint models can be used to estimate the forecast distribution of $\{S_t\}$. For example, the one-step-ahead forecast probability function for $\{S_t\}$ can be obtained as follows, for $s=0, 1, 2, \ldots$:

$$\begin{aligned}
\mathrm{P}\left(S_{T+1} = s \mid S^{(T)}, N^{(T)}\right) \\
= \sum_{n=s}^{\infty} \mathrm{P}\left(S_{T+1}=s \mid S^{(T)}, N^{(T)}, N_{T+1}=n\right) \\
\times \mathrm{P}\left(N_{T+1}=n \mid S^{(T)}, N^{(T)}\right) \\
= \sum_{n=s}^{\infty} \mathrm{P}\left(S_{T+1}=s \mid S^{(T)}, N_{T+1}=n\right) \\
\times \mathrm{P}\left(N_{T+1}=n \mid N^{(T)}\right).
\end{aligned}$$

Similarly, it is possible to give an expression for the k-step-ahead forecast distribution for $k > 1$, but this involves multiple summations which rapidly become computationally unmanageable as k increases. A practical method of estimating the forecast distribution is by means of Monte Carlo methods. One generates many realizations of $\{N_t,\ t = T+1, \ldots, T+k \mid N^{(T)}\}$ and of $\{S_t,\ t = T+1, \ldots, T+k \mid N^{(T+k)},\ S^{(T)}\}$. The k-step-ahead forecast distribution

$$\mathrm{P}\left(S_{T+k} = s_{T+k} \mid S^{(T)},\ N^{(T)}\right)$$

can then be estimated as the empirical distribution of the generated values S_{T+k}.

3.9 Discussion

Although many variations on a hidden Markov theme have been presented here, it will be clear that the list is not exhaustive. One could, for instance, consider models allowing for some, preferably simple, conditional dependence along time. Which further variations are worth pursuing will be determined by applications. As a general class of models for discrete-valued series, the hidden Markov models are certainly very flexible and able to accommodate the characteristics of many types of data.

The device of using an underlying Markov chain to introduce dependence between variables that are otherwise independent results in a parsimony of parametrization that is not easily achieved by competing models. Compare for instance second-order Markov chains on three states with binomial-hidden Markov models with

$n_t = 2$ for all t. (In all these models the observations take one of three values.) The general second-order Markov chain has 18 parameters and the corresponding Raftery model has seven. The hidden Markov model has m^2 parameters if its underlying Markov chain has m states. Hence even a hidden Markov model with four states has fewer parameters than does a general second-order Markov chain, and a hidden Markov model with two states has fewer parameters than the Raftery model.

Furthermore, the relative ease with which the likelihood may be evaluated and maximized with respect to the parameters greatly adds to the usefulness of hidden Markov models as practical statistical tools. The particular class of hidden Markov models we have chosen to concentrate on in this book, those based on a stationary Markov chain, has the added advantage that the theoretical autocorrelation function can be found and compared with the sample autocorrelations for the purpose of model verification — provided that the observations are quantitative and the autocorrelations therefore meaningful. While it is true that the EM algorithm is not as easily applied to hidden Markov models of this class as to those of Leroux and Puterman, the availability of sophisticated optimization software and the conceptual simplicity of direct numerical maximization mean that this is not really a disadvantage.

We now present in Chapter 4 a number of illustrative examples of applications of the models introduced in this book.

CHAPTER 4

Applications

4.1 Introduction

The purpose of this chapter is to describe the application of some of the models introduced in Chapters 2 and 3 to data from a variety of subjects. In the application of any new methodology it is helpful to have available both examples of its use which appear successful and examples which appear less so. We therefore report here not only those applications in which hidden Markov models have turned out to be useful, but also some in which the nature of such models seems to make them inappropriate, or at least less appropriate than some competing model. It is hoped that this will aid the reader in deciding whether a hidden Markov model is likely to be useful in a particular application.

All the hidden Markov models described in this chapter were fitted by numerical maximization of the likelihood. With the exception of the models appearing in section 4.4, the optimization was performed by means of the sequential quadratic programming routine E04UCF in the NAG Fortran Library (Numerical Algorithms Group, 1992). For the models of section 4.4 the Nelder–Mead simplex algorithm, as implemented by Press *et al.* (1986), was used instead. (This can be done in all cases where there are no constraints on the parameters other than possibly simple upper and lower bounds: see section 2.7.1.)

Throughout the chapter, k denotes the number of parameters estimated in order to fit a model, and l the log-likelihood. The model selection criteria used, AIC and BIC, are given by $-2l + 2k$ and $-2l + k \log T$ respectively, with T as usual denoting the length of the observation sequence. The use of these criteria is discussed in section 2.10.

Most of the data sets discussed are printed in Appendix B. It is intended that all the data, as well as the Fortran programs used to perform many of the analyses reported here, will be available by

4.2 The durations of successive eruptions of the Old Faithful geyser

Azzalini and Bowman (1990) have presented an interesting time series analysis of data on eruptions of the Old Faithful geyser in the Yellowstone National Park in the USA. The data consist of two series of length 299, collected continuously from 1 August to 15 August 1985. The first series is of the durations, d_t, of successive eruptions, and the second is of the waiting times, w_t, preceding those eruptions (defined as the differences between the starting times of the relevant eruptions).

It is true of both series that most of the observations can be described as either long or short, with very few observations intermediate in length, and with relatively low variation within the low and high groups. It is therefore natural to treat these series as binary time series: Azzalini and Bowman do so by discretizing the two series at 3 and 68 minutes respectively, denoting short by 0 and long by 1. (There is, in respect of the durations series, the complication that some of the eruptions were observed only as short, medium or long, and the medium durations have to be treated as either short or long. This point will be discussed further in due course.) It emerges that $\{D_t\}$ and $\{W_t\}$, the discretized versions of the series $\{d_t\}$ and $\{w_t\}$, are very similar — almost identical, in fact — and Azzalini and Bowman therefore concentrate on the series $\{D_t\}$ as representing most of the information relevant to the state of the system. We do the same here.

Table B.1 of Appendix B presents the series $\{D_t\}$, using 0 for short eruptions and 1 for long, and treating the two eruptions of medium length as long. On examination of the series $\{D_t\}$ one notices that 0 is always followed by 1, and 1 by either 0 or 1. A summary of the data is displayed in the 'observed no.' column of Table 4.2 on p. 141.

4.2.1 Markov chain models

In brief, what Azzalini and Bowman first did was to fit a (first-order) Markov chain model. This model seemed quite plausible from a geophysical point of view, but did not match the sample ACF at all well. They then fitted a second-order Markov chain

Table 4.1. *Geyser data: sample autocorrelation function* ($\hat{\rho}_k$) *and partial autocorrelation function* ($\hat{\phi}_{kk}$) *of the series* $\{D_t\}$ *of short and long eruptions.*

k	1	2	3	4	5	6	7	8
$\hat{\rho}_k$	−0.538	0.478	−0.346	0.318	−0.256	0.208	−0.161	0.136
$\hat{\phi}_{kk}$	−0.538	0.266	−0.021	0.075	−0.021	−0.009	0.010	0.006

model, which matched the ACF much better, but did not attempt a geophysical interpretation for this second model. We describe in some detail what Azzalini and Bowman did, and then fit hidden Markov models and compare them with their models.

With the convention that a medium is treated as a long, and using the estimator of the ACF described by Box and Jenkins (1976, pp. 32–33), Azzalini and Bowman estimated the ACF and PACF of $\{D_t\}$ as displayed in Table 4.1. Since the sample ACF is not even very approximately of the form α^k, a Markov chain is not a satisfactory model. (Although the high value of $\hat{\phi}_{22}$ also casts doubt on the adequacy of a Markov chain model, this is not really additional evidence. Since $\rho_k = \alpha^k$, for all positive integers k, implies that $\phi_{rr} = 0$ for $r > 1$ — see section 1.2.1 — the high value of $\hat{\phi}_{22}$ can be regarded as merely a restatement of the evidence provided by the ACF.) Azzalini and Bowman therefore fitted a second-order Markov chain, which turned out not to be consistent with a first-order model. They mention also that they fitted a third-order model, which did produce estimates consistent with a second-order model.

An estimate of the transition probability matrix of the first-order Markov chain, based on maximizing the likelihood conditional on the first observation, is

$$\begin{pmatrix} 0 & 1 \\ \frac{105}{194} & \frac{89}{194} \end{pmatrix} = \begin{pmatrix} 0 & 1 \\ 0.5412 & 0.4588 \end{pmatrix}. \quad (4.1)$$

Although it is not central to this discussion, it is worth noting that unconditional maximum likelihood is very easy in this case. Because there are no transitions from 0 to 0, an explicit formula applies: see Bisgaard and Travis (1991). The result is that the transition probability matrix is estimated as

$$\begin{pmatrix} 0 & 1 \\ 0.5404 & 0.4596 \end{pmatrix}. \quad (4.2)$$

This serves to confirm as reasonable the expectation that, for a series of length 299, estimation by conditional maximum likelihood differs very little from unconditional.

The estimate of the t.p.m. reported by Azzalini and Bowman is slightly different from (4.1), however: $\begin{pmatrix} 0 & 1 \\ 0.557 & 0.443 \end{pmatrix}$. This appears to arise as the matrix $\begin{pmatrix} 0 & 1 \\ \frac{107}{192} & \frac{85}{192} \end{pmatrix}$, which is in turn based on the convention that a medium is treated as a *short* (and on conditional maximum likelihood). Azzalini and Bowman therefore seem to have used one convention when estimating the ACF, and the other when estimating the t.p.m. A similar comment applies to their second-order model. We use the convention throughout that a medium is treated as a long.

Since the sequence (0,0) does not occur, the three states needed to express the second-order model as a first-order Markov chain are, in order: (0,1), (1,0), (1,1). The corresponding t.p.m. is:

$$\begin{pmatrix} 0 & \frac{69}{104} & \frac{35}{104} \\ 1 & 0 & 0 \\ 0 & \frac{35}{89} & \frac{54}{89} \end{pmatrix} = \begin{pmatrix} 0 & 0.6635 & 0.3365 \\ 1 & 0 & 0 \\ 0 & 0.3933 & 0.6067 \end{pmatrix}. \quad (4.3)$$

In view of the minor discrepancy noted above, the theoretical ACF we get for the second-order model differs from that quoted by Azzalini and Bowman. The ACF of model (4.3), which has stationary distribution $\frac{1}{297}(104, 104, 89)$, can be computed from

$$\begin{aligned} \rho_k &= \frac{\mathrm{E}(D_t D_{t+k}) - \mathrm{E}(D_t)\mathrm{E}(D_{t+k})}{\mathrm{Var}(D_t)} \\ &= \frac{297^2 \mathrm{P}(D_t = D_{t+k} = 1) - 193^2}{193 \times 104}. \end{aligned}$$

The resulting figures for $\{\rho_k\}$ are given in Table 4.3 on p. 142, and match the sample ACF $\{\hat\rho_k\}$ well, slightly better (as one would expect) than do Azzalini and Bowman's own figures for $\{\rho_k\}$.

4.2.2 Hidden Markov models

We now discuss the use of hidden Markov models for the series $\{D_t\}$. Binomial-hidden Markov models with $n_t = 1$ and $m = 1, 2, 3$

Table 4.2. *Geyser data: observed numbers of short and long eruptions and various transitions, compared with those expected under the two-state hidden Markov model.*

	observed no.	expected no.
short eruptions (0)	105	105.0
long eruptions (1)	194	194.0
Transitions:		
from 0 to 0	0	0.0
from 0 to 1	104	104.0
from 1 to 0	105	104.9
from 1 to 1	89	89.1
from (0,1) to 0	69	66.7
from (0,1) to 1	35	37.3
from (1,0) to 1	104	104.0
from (1,1) to 0	35	37.6
from (1,1) to 1	54	51.4

or 4 were fitted to the series $\{D_t\}$. The notation used here is as in section 2.3.

We describe the two-state model in some detail. This model has log-likelihood -127.31, $\Gamma = \begin{pmatrix} 0.000 & 1.000 \\ 0.827 & 0.173 \end{pmatrix}$ and $p = (0.225, 1.000)$. That is, there are two (unobserved) states, state 1 always being followed by state 2, and state 2 by state 1 with probability 0.827. In state 1 a long has probability 0.225, in state 2 it has probability 1. A convenient interpretation of this model is that it is a rather special stationary two-state Markov chain, with some noise present in the first state: if the probability 0.225 were instead zero, the model would be exactly a Markov chain. A long has unconditional probability $\delta\lambda(1)\mathbf{1}' = 0.649$. A long is followed by a short with probability $\delta\lambda(1)\Gamma\lambda(0)\mathbf{1}'/\delta\lambda(1)\mathbf{1}' = 0.541$, and a short is always followed by a long. A comparison of observed numbers of zeros, ones and transitions, with the numbers expected under this model, is presented in Table 4.2. (A similar comparison in respect of the second-order Markov chain model would not be informative, because in that case parameters have been estimated by a method which forces equality of observed and expected numbers of first- and second-order transitions.)

In the notation of section 2.4.2, the ACF is given for all $k \in \mathbf{N}$ by $\rho_k = (1+\alpha)^{-1}w^k$, where $w = -0.827$ and $\alpha = 0.529$. Hence

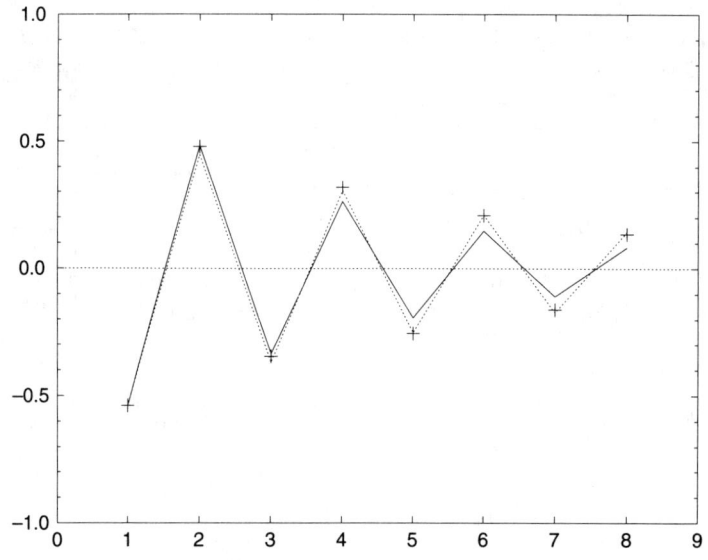

Figure 4.1. *Geyser data: sample autocorrelation function, and ACF of two models. The crosses (+) denote the sample ACF, the continuous line the ACF of the second-order Markov chain (model (4.3)), and the dotted line that of the two-state hidden Markov model.*

Table 4.3. *Geyser data: sample ACF compared with the ACF of the second-order Markov chain and the hidden Markov model.*

k	1	2	3	4	5	6	7	8
ρ_k for model (4.3)	−0.539	0.482	−0.335	0.262	−0.194	0.147	−0.110	0.083
sample ACF, $\hat{\rho}_k$	−0.538	0.478	−0.346	0.318	−0.256	0.208	−0.161	0.136
ρ_k for HM model	−0.541	0.447	−0.370	0.306	−0.253	0.209	−0.173	0.143

$\rho_k = 0.654 \times (-0.827)^k$. In Figure 4.1 and Table 4.3 the resulting figures are compared with the sample ACF and with the theoretical ACF of the (corrected) second-order Markov chain model, i.e. model (4.3).

It seems reasonable to conclude that the hidden Markov model fits the sample ACF well: not quite as well as the second-order Markov chain model as regards the first three autocorrelations, but better for longer lags.

Table 4.4. *Geyser data: percentiles of bootstrap sample of estimators of parameters of two-state hidden Markov model.*

percentile:	5th	25th	median	75th	95th
$\hat{\gamma}_2$	0.709	0.793	0.828	0.856	0.886
$\hat{p}_1$	0.139	0.191	0.218	0.244	0.273

The parametric bootstrap, with a sample size of 100, was used to estimate the means and covariances of the maximum likelihood estimators of the four parameters γ_1, γ_2, p_1 and p_2. That is, 100 series of length 299 were generated from the two-state hidden Markov model described above, and a model of the same type fitted in the usual way to each of these series. The random number generator used was that of Wichmann and Hill (1982). The sample mean vector for the four parameters is (1.000, 0.819, 0.215, 1.000), and the sample covariance matrix is:

$$\begin{pmatrix} 0 & 0 & 0 & 0 \\ 0 & 0.003303 & 0.001540 & 0 \\ 0 & 0.001540 & 0.002065 & 0 \\ 0 & 0 & 0 & 0 \end{pmatrix}.$$

The estimated standard deviations of the estimators are therefore (0.000, 0.057, 0.045, 0.000). (The zero standard errors are of course not typical: they are a consequence of the rather special nature of the model from which we are generating the series.) As a further indication of the behaviour of the estimators we present in Table 4.4 selected percentiles of the bootstrap sample of values of $\hat{\gamma}_2$ and $\hat{p}_1$. From these bootstrap results it appears that, for this application, the maximum likelihood estimators have fairly small standard deviations and are not markedly asymmetric. It should, however, be borne in mind that the estimate of the distribution of the estimators which is provided by the parametric bootstrap is derived under the assumption that the model fitted to the data is correct.

There is, however, a further class of models which generalizes both the two-state second-order Markov chain and the two-state hidden Markov model as described above. This is the class of two-state second-order hidden Markov models on state space $\{0, 1\}$. Such models (*inter alia*) are described in some detail in section 3.2. By using the recursion (3.3) for the probability $\nu_t(i, j; s_1, \ldots, s_T)$, with the appropriate scaling, it is almost as straightforward to compute the likelihood of a second-order model as a first-order one and to fit models by maximum likelihood. In the present example the resulting probabilities of a long eruption are 0.072 (state 1) and 1.000 (state 2). The underlying process (or parameter process) is a two-state second-order Markov chain with associated first-order Markov chain having transition probability matrix

$$\begin{pmatrix} 1-a & a & 0 & 0 \\ 0 & 0 & 0.717 & 0.283 \\ 0 & 1 & 0 & 0 \\ 0 & 0 & 0.441 & 0.559 \end{pmatrix}. \qquad (4.4)$$

Here a may be taken to be any real number between 0 and 1, and the four states used for this purpose are, in order: (1,1), (1,2), (2,1), (2,2). The log-likelihood is -126.90. (Clearly the state (1,1) can be disregarded above without loss of information, in which case the first row and first column are deleted from the matrix (4.4).)

It should be noted that the second-order Markov chain used here as underlying process is the general four-parameter model, not the Pegram–Raftery submodel, which has three parameters. From the comparison which follows it will be seen that a hidden Markov model based on a Pegram–Raftery second-order chain is in this case not worth pursuing, because with a total of five parameters it cannot produce a log-likelihood value better than -126.90. (The two four-parameter models fitted produce values of -127.31 and -127.12, which by AIC and BIC are preferable to a log-likelihood of -126.90 for a five-parameter model.)

4.2.3 Comparison of models

We now compare all the models considered on the basis of their unconditional log-likelihoods, denoted by l, and AIC and BIC. For instance, in the case of model (4.1), the first-order Markov chain fitted by conditional maximum likelihood, we have:

$$l = \log(194/299) + 105\log(105/194) + 89\log(89/194) = -134.2426.$$

The comparable figure for model (4.2) is -134.2423: in view of the minute difference we shall here ignore the distinction between estimation by conditional and by unconditional maximum likelihood. For the second-order Markov chain model (4.3), l is given by

$$\log(104/297) + 35\log(35/104) + 69\log(69/104)$$
$$+ 35\log(35/89) + 54\log(54/89),$$

Table 4.5. *Geyser data: comparison of models on the basis of AIC and BIC.*

model	k	$-l$	AIC	BIC
1-state hidden Markov (i.e. independence)	1	193.80	389.60	393.31
Markov chain	2	134.24	272.48	279.88
second-order Markov chain	4	127.12	**262.24**	**277.04**
2-state hidden Markov	4	127.31	262.62	277.42
3-state hidden Markov	9	126.85	271.70	305.00
4-state hidden Markov	16	126.59	285.18	344.39
2-state second-order HM	6	126.90	265.80	288.00

and equals -127.12. Table 4.5 presents a comparison of seven types of model, including for completeness the one-state hidden Markov model, i.e. the model which assumes independence of the consecutive observations. (Given the strong negative dependence apparent in the data, it is not surprising that this model is so much inferior to the others considered.)

From the table it emerges that, on the basis of AIC and BIC, only the second-order Markov chain and the two-state (first-order) hidden Markov model are worth considering. In the comparison, both of these models are taken to have four parameters, because, although the observations suggest that the sequence (short, short) cannot occur, there is no *a priori* reason to make such a restriction.

While it is true that the second-order Markov chain seems a slightly better model on the basis of the model selection exercise described above, and possibly on the basis of the ACF, both are reasonable models capable of describing the principal features of the data without using an excessive number of parameters. The hidden Markov model perhaps has the advantage of relative simplicity, given its nature as a Markov chain with some noise in one of the states. Azzalini and Bowman note that their second-order Markov chain model would require a more sophisticated interpretation than does their first-order model. Either a longer series of observations or a convincing geophysical interpretation for one model rather than the other would be needed to take the discussion further.

4.2.4 Forecast distributions

We conclude this analysis by demonstrating how the two-state hidden Markov model may be used to provide forecasts. As it happens, the last observation in the series $\{D_t\}$ (the 299th) is zero, so that with probability one the next one is 1. We therefore give here the two-step-ahead and joint two- and three-step-ahead forecast distributions implied by the model. (Higher-order joint distributions and forecasts further into the future involve no essentially different features.) The relevant probabilities are given by ratios of likelihood values, as described in section 2.6.

Given the full history, the probability that $D_{301}=1$ (and $D_{300}=1$) is 0.359. The probabilities that, given the history, $D_{301}=i$ and $D_{302}=j$ (and $D_{300}=1$) are given in Table 4.6. For comparison we state also the corresponding figures for the second-order Markov chain model (4.3). The conditional probability that $D_{301}=1$ is in that case 0.337, and the joint forecast distribution of D_{301} and D_{302} is also given in Table 4.6.

Table 4.6. *Geyser data: joint two- and three-step-ahead forecast distributions for the two-state hidden Markov model (left) and the second-order Markov chain model (right).*

	$j=0$	1		$j=0$	1
$i=0$	0.000	0.641	$i=0$	0.000	0.663
1	0.111	0.248	1	0.132	0.204

4.3 Epileptic seizure counts

Albert (1991) and Le, Leroux and Puterman (1992) describe the fitting of two-state Poisson-hidden Markov models to series of daily counts of epileptic seizures in one patient. Such models appear to be a promising tool for the analysis of seizure counts, the more so as there are suggestions in the neurology literature that the susceptibility of a patient to seizures may vary in a fashion that can reasonably be represented by a Markov chain: see Hopkins, Davies and Dobson (1985). Another promising approach, that of Franke and Seligmann (1993), is to use an AR(1) analogue based on thinning: see section 1.5 for a discussion of models based on thinning.

In this section we apply our methods to a corrected version of the series analysed by Le et al., which appears as Table B.2 in our Appendix B. The original series of observations is 225 days long, and consists of the numbers of myoclonic seizures suffered by a patient on each of those days. However, we have been informed (Le, personal communication) that there is an error in the data as published, and that observations 92–112 inclusive should be deleted. The corrected series is therefore 204 observations long, and, unless we indicate otherwise, what we present here refers to that corrected series.

Le et al. use a model of the type described by Leroux and Puterman (1992) and discussed in section 2.7. That is, the model does not assume that the underlying Markov chain is stationary. The likelihood maximized is conditional on the Markov chain starting in a particular state, that state in fact which yields the higher value of the maximized likelihood. We discuss similar models here, but based on a stationary Markov chain and fitted by maximization of the unconditional likelihood of the observations. We consider models with from one to four underlying states, and note that both AIC and BIC choose the two-state model; we find the ACF of that model and compare it with the sample ACF; we compare the marginal properties of the model with those of the data; and we use the pseudo-residuals and forecast pseudo-residuals defined in section 2.8 to check for outliers in the series of counts.

Table 4.7. *Epileptic seizure counts: comparison of several stationary hidden Markov models by means of AIC and BIC.*

no. of states	k	l	AIC	BIC
1	1	−232.15	466.31	469.63
2	4	−211.68	**431.36**	**444.64**
3	9	−208.45	434.90	464.76
4	16	−201.68	435.36	488.45

Table 4.7 uses AIC and BIC to compare models based on a stationary Markov chain with one, two, three or four states. (The one-state model is just the model which assumes that the observations are realizations of independent Poisson random variables with a common mean. That mean is the only parameter.) From the table we see that, of the four models considered, that with

two states is chosen both by AIC and (by a large margin) by BIC. We therefore do not consider the other models further. The details of the two-state model are as follows. The Markov chain has transition probability matrix

$$\begin{pmatrix} 0.965 & 0.035 \\ 0.027 & 0.973 \end{pmatrix},$$

and starts from the stationary distribution (0.433, 0.567). The seizure frequencies in states 1 and 2 are 1.167 and 0.262 respectively.

Table 4.8. *Sample ACF for the epileptic seizure counts.*

k	1	2	3	4	5	6	7	8
$\hat{\rho}_k$	0.236	0.201	0.199	0.250	0.157	0.181	0.230	0.242

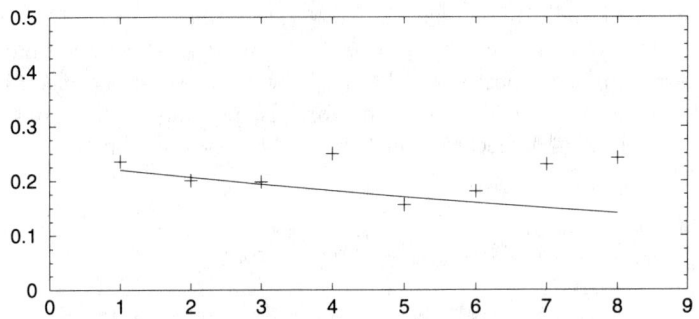

Figure 4.2. *Sample and theoretical ACF for the epileptic seizures data. The crosses denote the sample ACF, and the continuous line the theoretical ACF of a stationary two-state Poisson-hidden Markov model based on 204 observations.*

The ACF of the model can be computed by the methods of section 2.4.1. In the notation used there it is given, for all positive integers k, by

$$\begin{aligned} \rho_k &= \left(1 + \frac{\delta\lambda'}{(\lambda_2 - \lambda_1)^2 \delta_1 \delta_2}\right)^{-1} w^k \\ &= 0.235 \times 0.939^k. \end{aligned}$$

EPILEPTIC SEIZURE COUNTS

Table 4.9. *Observed and expected numbers of days with* $r = 0, 1, 2, \ldots$ *epileptic seizures.*

r	observed no.	expected no.
0	117	116.5
1	54	55.4
2	23	21.8
3	7	7.5
4	2	2.1
5	0	0.5
≥ 6	1	0.1
	204	203.9

Table 4.8 gives the corresponding sample ACF. Figure 4.2, which displays both, shows that the agreement between sample and theoretical ACF is reasonably close.

The marginal properties of the model can be assessed from Table 4.9, which gives the observed and expected numbers of days on which there were 0, 1, 2, ..., 6 or more seizures. Agreement is excellent.

We now use the techniques of section 2.8 to check for outliers under the two-state model we have chosen. Figure 4.3 is a pseudo-residual plot, and from it we see that three observations out of the 204 stand out as extreme, namely those at times 106, 147 and 175. They all yield upper-tail p-values less than 0.01: respectively 0.00130, 0.00325, and 0.00038.

A Bonferroni lower bound for the probability that all 204 p-values will fall within the middle 99.92% of their respective conditional distributions is $1 - 204 \times 0.0008 = 0.84$, and the corresponding figure based on independence is 0.85. This suggests that observation 175 is sufficiently extreme (relative to its neighbours) to be considered an outlier. As regards observation 106, the lower bound for the probability that all 204 p-values lie within the middle 99.7% of their distributions is 0.39, and the figure based on independence is 0.54. There is no strong evidence, therefore, for considering observation 106 to be an outlier relative to its neighbours.

It is interesting to note that observations 23 and 175, both of which represent four seizures in one day, yield rather different

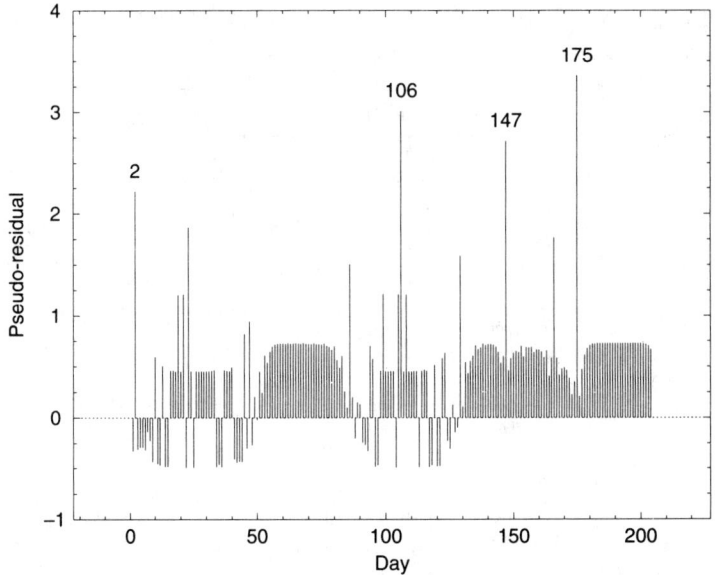

Figure 4.3. *Epileptic seizures data: pseudo-residuals relative to stationary two-state hidden Markov model fitted to all 204 observations.*

(upper-tail) p-values: 0.03077 and 0.00038 respectively. The reason for this is clear when one notes that most of the near neighbours (in time) of observation 175 are zero, which is not true of the neighbours of observation 23. Observation 23 is much less extreme relative to its neighbours than is 175, and this is reflected in the p-value. Similarly, observation 106 (six seizures in a day) is less extreme relative to its neighbours than is observation 175 (four seizures) relative to its neighbours.

However, a more interesting exercise is to see whether, if a model had been fitted from (say) the first 100 observations only, day-by-day monitoring thereafter by means of forecast pseudo-residuals would have identified any outliers. The two-state model fitted from the first 100 observations has transition probability matrix

$$\begin{pmatrix} 0.983 & 0.017 \\ 0.042 & 0.958 \end{pmatrix},$$

and seizure frequencies 1.049 and 0.258.

From a plot of forecast pseudo-residuals (Figure 4.4) we see that the same three observations stand out: days 106, 147 and 175, with

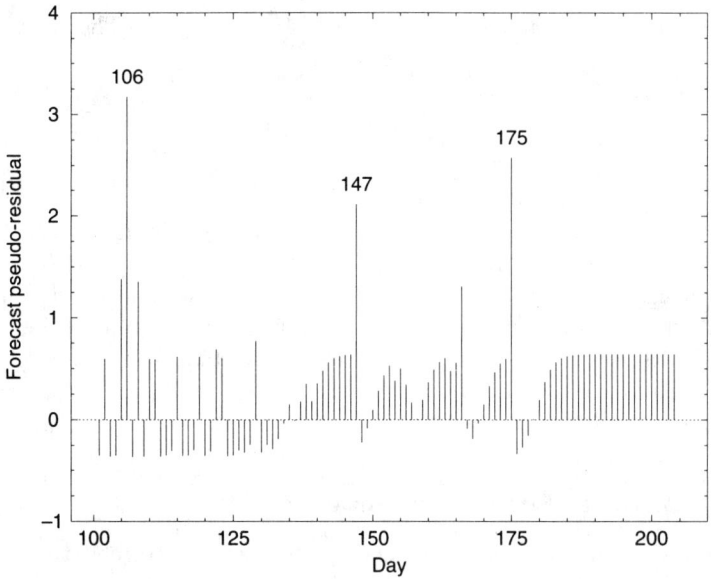

Figure 4.4. *Epileptic seizures data: forecast pseudo-residuals relative to stationary two-state hidden Markov model fitted to data for days 1–100.*

corresponding forecast p-values of 0.00074, 0.01716 and 0.00493. The lower bound for the probability that all 104 values lie in the middle 99.85% of their distributions is 0.84; the corresponding figure under the assumption of independence is 0.86. For 99% the corresponding figures are 0 and 0.35. Only observation 106 can therefore be said to emerge from such a monitoring procedure as an outlier relative to its predecessors.

In passing, it is perhaps worth noting that it was the use of pseudo-residuals that first alerted us to the possibility that the data as originally published might contain an error. The pseudo-residuals (and p-values) for observations 106 and 127, although extreme, were equal, which seemed surprising until inspection of the data revealed that, in the original data, observations 92–112 and 113–133 were identical.

Finally, and purely for the purposes of comparison, we present the model fitted by Le *et al.* to the original data, and the corresponding model based on a stationary Markov chain. The model fitted by Le *et al.* is as follows. The Markov chain, on states which we shall refer to as 1 and 2, starts in state 1 and has transition probability matrix given by $\begin{pmatrix} 0.976 & 0.024 \\ 0.014 & 0.986 \end{pmatrix}$. The seizure frequencies are 1.255 and 0.287 in states 1 and 2 respectively. The log-likelihood value achieved is (apart from constant terms) -189.35. The corresponding stationary model has transition probability matrix $\begin{pmatrix} 0.976 & 0.024 \\ 0.020 & 0.980 \end{pmatrix}$, starts from the stationary distribution (0.449, 0.551) of the Markov chain, and has seizure frequencies 1.270 and 0.288 respectively. The log-likelihood is -246.73 or, if constant terms are omitted, -189.89. As one would expect, this last value is slightly inferior to the corresponding value of Le *et al.* (-189.35): they maximize over one more (binary) variable, the initial state of the Markov chain. The parameter values in the two models do not differ greatly.

4.4 Births at Edendale hospital

Haines, Munoz and van Gelderen (1989) have described the fitting of Gaussian ARIMA models to various discrete-valued time series related to births occurring during a 16-year period at Edendale Hospital in Natal, South Africa. The data include (*inter alia*) monthly totals of mothers delivered and deliveries by various methods at the Obstetrics Unit of that hospital in the period from February 1970 to January 1986 inclusive. Although 16 years of data were available, Haines *et al.* considered only the final eight years' observations when modelling any series other than the total deliveries. This was because, for some of the series they modelled, the model structure was found not to be stable over the full 16-year period. In this analysis we have in general modelled only the last eight years' observations.

4.4.1 Models for the proportion Caesarean

One of the series considered by Haines *et al.*, to which they fitted two models, was the number of deliveries by Caesarean section. From their models they drew the conclusions (in respect of this particular series) that there is a clear dependence of present on past observations, and that there is a clear linear upward trend. In this section (4.4.1) we describe the fitting of (discrete-valued) Markov regression and hidden Markov models to this series. These models are of course rather different from those fitted by Haines *et al.* in that the latter, being based on the normal distribution, are continuous-valued. Furthermore, the discrete-valued models make it possible to model the proportion (as opposed to the number) of Caesareans performed in each month. Of the models proposed here, one type is observation-driven and the other parameter-driven. The most important conclusion drawn from the discrete-valued models, and one which the Gaussian ARIMA models did not provide, is that there is a strong upward time trend in the proportion of the deliveries that are by Caesarean section.

The two models which Haines et al. fitted to the time series of Caesareans performed, and which they found to fit very well, may be described as follows. Let Z_t denote the number of Caesareans in month t (February 1978 being month 1), and let the process $\{a_t\}$ be Gaussian white noise, i.e. uncorrelated random shocks distributed normally with zero mean and common variance σ_a^2. The first model fitted is the ARIMA(0,1,2) model with constant term:

$$\nabla Z_t = \mu + a_t - \theta_1 a_{t-1} - \theta_2 a_{t-2}. \tag{4.5}$$

The maximum likelihood estimates of the parameters, with associated standard errors, are: $\hat{\mu} = 1.02 \pm 0.39, \hat{\theta}_1 = 0.443 \pm 0.097, \hat{\theta}_2 = 0.393 \pm 0.097$ and $\hat{\sigma}_a^2 = 449.25$. The second model is an AR(1) with linear trend:

$$Z_t = \beta_0 + \beta_1 t + \phi Z_{t-1} + a_t, \tag{4.6}$$

with parameter estimates as follows: $\hat{\beta}_0 = 120.2 \pm 8.2, \hat{\beta}_1 = 1.14 \pm 0.15, \hat{\phi} = 0.493 \pm 0.092$ and $\hat{\sigma}_a^2 = 426.52$.

Both of these models, (4.5) and (4.6), provide support for the conclusion of Haines et al. that there is a dependence of present on past observations, and a linear upward trend. Furthermore, the models are nonseasonal: the Box–Jenkins methodology used found no seasonality in the Caesareans series. The X-11-ARIMA seasonal adjustment method employed in an earlier study (Munoz, Haines and van Gelderen, 1987) did, however, find some evidence, albeit weak, of a seasonal pattern in the Caesareans series similar to a pattern that was observed in the 'total deliveries' series. (This latter series shows marked seasonality, with a peak in September, and in Haines et al. (1989) is modelled by the seasonal ARIMA model $(0, 1, 1) \times (0, 1, 1)_{12}$.)

It is of some interest to model the proportion, rather than the number, of Caesareans in each month. (See Figure 4.5 for a plot of this proportion for the years 1978–1986.)

It could be the case, for instance, that any trend, dependence or seasonality apparently present in the number of Caesareans is

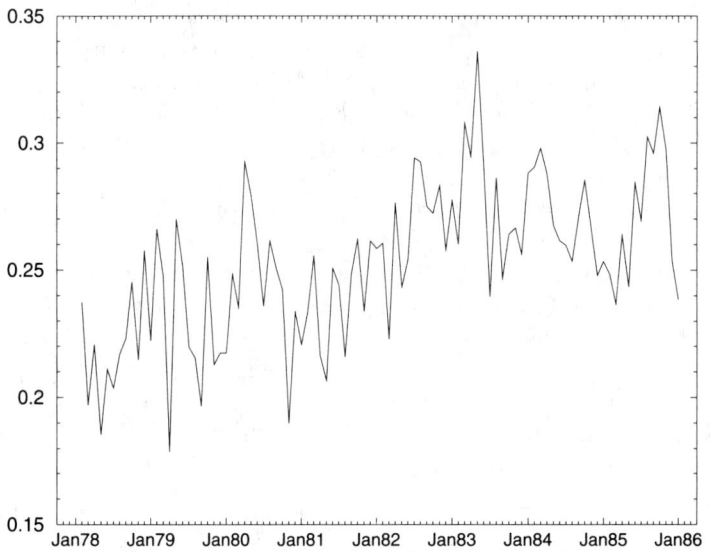

Figure 4.5. *Edendale births: monthly numbers of deliveries by Caesarean section, as a proportion of all deliveries, February 1978 – January 1986.*

largely inherited from the total deliveries, and a constant proportion Caesarean is an adequate model. On the other hand, it could be the case that there is an upward trend in the proportion of the deliveries that are by Caesarean and this accounts at least partially for the upward trend in the number of Caesareans. The two classes of model that we discuss in this section condition on the total number of deliveries in each month and seek to describe the principal features of the proportion Caesarean.

Now let n_t denote the total number of deliveries in month t. A very general possible model for $\{Z_t\}$ which allows for trend, dependence on previous observations and seasonality, in the proportion Caesarean, is as follows. Suppose that, conditional on the history $Z^{(t-1)} = \{Z_s : s \leq t-1\}$, Z_t is distributed binomially with parameters n_t and p_t, where, for some positive integer q,

$$\begin{aligned}\operatorname{logit} p_t &= \alpha_1 + \alpha_2 t + \beta_1(Z_{t-1}/n_{t-1}) + \beta_2(Z_{t-2}/n_{t-2}) + \cdots \\ &\quad + \beta_q(Z_{t-q}/n_{t-q}) + \gamma_1 \sin(2\pi t/12) + \gamma_2 \cos(2\pi t/12).\end{aligned}$$

(4.7)

This is a Markov regression model and generalizes the model described by Cox (1981) as an 'observation-driven linear logistic autoregression', in that it incorporates trend and seasonality and is based on a binomial rather than a Bernoulli distribution. Clearly it is possible to add further terms to the above expression for logit p_t to allow for the effect of any further covariates thought relevant, e.g. the number or proportion of deliveries by various instrumental techniques.

It does not seem possible to formulate an unconditional maximum likelihood procedure to estimate the parameters α_1, α_2, β_1, ..., β_q, γ_1 and γ_2 of the model (4.7). It is, however, straightforward to compute estimates of these parameters by using a program such as GLIM, GENSTAT, S-PLUS or SAS to maximize a conditional likelihood. If for instance no observations earlier than Z_{t-1} appear in the model, the product

$$\prod_{t=1}^{96} \binom{n_t}{z_t} p_t^{z_t}(1-p_t)^{n_t-z_t}$$

is the likelihood of $\{Z_t : t = 1, \ldots, 96\}$, conditional on Z_0. Maximization thereof with respect to $\alpha_1, \alpha_2, \beta_1, \gamma_1$ and γ_2 yields estimates of these parameters, and can be accomplished simply by performing a logistic regression of Z_t on t, Z_{t-1}/n_{t-1}, $\sin(2\pi t/12)$ and $\cos(2\pi t/12)$.

In the search for a suitable model the following explanatory variables were considered: t (i.e. the time in months, with February 1978 as month 1), t^2, the proportion Caesarean lagged one, two, three or twelve months, sinusoidal terms at the annual frequency (as in (4.7)), the calendar month, the proportion and number of deliveries by forceps or vacuum extraction, and the proportion and number of breech births. Model selection was performed by means of BIC. GLIM, the program used, does not provide l, the log-likelihood, but it does provide the (scaled) deviance, which is twice the difference between the maximum log-likelihood achievable in a full model and that achieved in the model under investigation (McCullagh and Nelder, 1989, p. 33). Here the maximum log-likelihood of a full model is -321.04, from which it follows that $-l = 321.04 + \frac{1}{2} \times$ deviance. The best models found with between one and four explanatory variables (other than the constant term) are listed in Table 4.10, as well as several other models that may be of interest. These other models are: the model with constant term only; the model with Z_{t-1}/n_{t-1}, the previous propor-

Table 4.10. *Edendale births: models fitted by GLIM to the logit of the proportion Caesarean.*

explanatory variables	coefficients	deviance	BIC
constant	−1.253	208.92	860.13
t (time in months)	0.003439		
constant	−1.594	191.70	847.47
t	0.002372		
previous proportion Caesarean	1.554		
constant	−1.445	183.63	843.97
t	0.001536		
previous proportion Caesarean	1.409		
no. forceps deliveries in month t	−0.002208		
constant	−1.446	175.60	**840.50**
t	0.001422		
previous proportion Caesarean	1.431		
no. forceps deliveries in month t	−0.002393		
October indicator	0.08962		
constant	−1.073	324.05	970.69
constant	−1.813	224.89	876.10
previous proportion Caesarean	2.899		
constant	−1.505	188.32	848.66
t	0.002060		
previous proportion Caesarean	1.561		
proportion instrumental deliveries in month t	−0.7507		
constant	−1.528	182.36	847.26
t	0.002056		
previous proportion Caesarean	1.590		
proportion instrumental deliveries in month t	−0.6654		
October indicator	0.07721		

tion Caesarean, as the only explanatory variable; and two models which replace the number of forceps deliveries as covariate by the proportion of instrumental deliveries. The last two models were included because the proportion of instrumental deliveries may seem a more sensible explanatory variable than the one it replaces: it will be observed, however, that the models involving the number of forceps deliveries are preferred by BIC. (By 'instrumental deliveries' we mean those which are either by forceps or by vacuum extraction.) BIC indicated that very little would be gained by inclusion of a fifth explanatory variable in the model.

The strongest conclusion we may draw from these models is that there is indeed a marked upward time trend in the proportion Caesarean. Secondly, there is positive dependence on the proportion Caesarean in the previous month. The negative association with the number (or proportion) of forceps deliveries is not surprising in view of the fact that delivery by Caesarean and by forceps are in some circumstances alternative techniques. As regards seasonality, the only possible seasonal pattern found in the proportion Caesarean is the positive 'October effect'. Among the calendar months only October stood out as having some explanatory power. As can be seen from Table 4.10, the indicator variable specifying whether the month was October was included in the 'best' set of four explanatory variables. A possible reason for an October effect is as follows. An overdue mother is more likely than others to give birth by Caesarean, and the proportion of overdue mothers may well be highest in October because the peak in total deliveries occurs in September.

Since the main conclusion emerging from the above logistic-linear models is that there is a marked upward time trend in the proportion Caesarean, it is of interest also to fit hidden Markov models with and without time trend. The hidden Markov models we use in this application have two states and are defined as follows. Suppose $\{C_t\}$ is a stationary homogeneous Markov chain on state space $\{1, 2\}$, with transition probability matrix

$$\Gamma = \begin{pmatrix} 1 - \gamma_1 & \gamma_1 \\ \gamma_2 & 1 - \gamma_2 \end{pmatrix}.$$

Suppose also that, conditional on the Markov chain, Z_t has a binomial distribution with parameters n_t and p_i, where $C_t = i$. A model without time trend assumes that p_1 and p_2 are constants and has four parameters. One possible model which allows p_i to depend on t has logit $_t p_i = \alpha_i + \beta t$ and has five parameters. A more general model yet, with six parameters, has logit $_t p_i = \alpha_i + \beta_i t$.

Maximization of the likelihood of the last eight years' observations, subject to the bounds of 0 and 1 on the probabilities involved, was performed by the Nelder–Mead simplex method (Press et al., 1986). The three resulting models, along with the associated log-

Table 4.11. *Edendale births: the three two-state hidden Markov models fitted to the proportion Caesarean. (The time in months is denoted by t, and February 1978 is month 1.)*

logit $_t p_i$	$-l$	BIC	γ_1	γ_2	α_1	α_2	β_1	β_2
α_i	420.322	858.90	0.059	0.086	-1.184	-0.9601	–	–
$\alpha_i + \beta t$	402.349	**827.52**	0.162	0.262	-1.317	-1.140	0.003297	
$\alpha_i + \beta_i t$	402.314	832.01	0.161	0.257	-1.315	-1.150	0.003249	0.003467

likelihood and BIC values, appear in Table 4.11. It may be seen that, of the three models, that with a single time-trend parameter and a total of five parameters achieves the smallest BIC value.

In detail, that model is as follows, t being the time in months and February 1978 being month 1:

$$\Gamma = \begin{pmatrix} 0.838 & 0.162 \\ 0.262 & 0.738 \end{pmatrix},$$

$$\text{logit } _t p_1 = -1.317 + 0.003297t,$$

$$\text{logit } _t p_2 = -1.140 + 0.003297t.$$

The model can therefore be described as consisting of a Markov chain with two fairly persistent states, along with their associated time-dependent probabilities of delivery being by Caesarean, the (upward) time trend being the same, on a logit scale, for the two states. State 1 is rather more likely than state 2, because the stationary distribution is (0.618, 0.382), and has associated with it a lower probability of delivery being by Caesarean. For state 1 that probability increases from 0.212 in month 1 to 0.269 in month 96, and for state 2 from 0.243 to 0.305. The corresponding unconditional probability increases from 0.224 to 0.283. It may or may not be possible to interpret the states as (for instance) nonbusy and busy periods in the Obstetrics Unit of the hospital, but without further information, e.g. on staffing levels, such an interpretation would be speculative.

It is true, however, that other models can reasonably be considered. One possibility, suggested by inspection of Figure 4.5, is that the proportion Caesarean was constant until January 1981, then increased linearly to a new level in about January 1983. Although we do not pursue such a model here, it is possible to fit a hidden Markov model incorporating this feature. (Models with change-points are discussed and used in section 4.10.)

If one wishes to use the chosen model to forecast the proportion Caesarean at time 97 for a given number of deliveries, what is needed is the one-step-ahead forecast distribution of Z_{97}, i.e. the distribution of Z_{97} conditional on $Z_1, \ldots, Z_{96}$. This is given by the likelihood of $Z_1, \ldots, Z_{97}$ divided by that of $Z_1, \ldots, Z_{96}$. More generally, the the k-step-ahead forecast distribution, i.e. the conditional probability that $Z_{96+k} = z$, is given by a ratio of likelihoods, as described in section 2.6.1.

The difference in likelihood between the hidden Markov models with and without time trend is convincing evidence of an upward trend in the proportion Caesarean, and confirms the main conclusion drawn above from the logistic-linear models. Although Haines et al. concluded that there is an upward trend in the number of Caesareans, it does not seem possible to draw any conclusion about the proportion Caesarean from their models, or from any other ARIMA models.

It is of interest also to compare the fit of the five-parameter hidden Markov model to the data with that of the logistic autoregressive models. Here it should be noted that the hidden Markov model produces a lower value of BIC (827.52) than does the logistic autoregressive model with four explanatory variables (840.50), even without making use of the additional information used by the logistic autoregression. It does not use z_0, nor does it use information on forceps deliveries or the calendar month. It seems therefore that hidden Markov models have considerable potential as simple yet flexible models for examining dependence on covariates (such as time) in the presence of serial dependence.

4.4.2 Models for the total number of deliveries

If one wishes to project the number of Caseareans, however, a model for the proportion Caesarean is not sufficient: one needs also a model for the total number of deliveries, which is a series of unbounded counts. The model of Haines et al. for the total deliveries was the seasonal ARIMA model $(0, 1, 1) \times (0, 1, 1)_{12}$ without constant term. For this series (unlike the others) they used all 16 years' data to fit the model, but we continue here to model only the final eight years' observations.

An attempt was first made to model the monthly totals of deliveries (depicted in Figure 4.6) by means of two-state Poisson-hidden Markov models. Two such models were considered: one with a single linear trend in the log of the conditional mean, i.e. a model

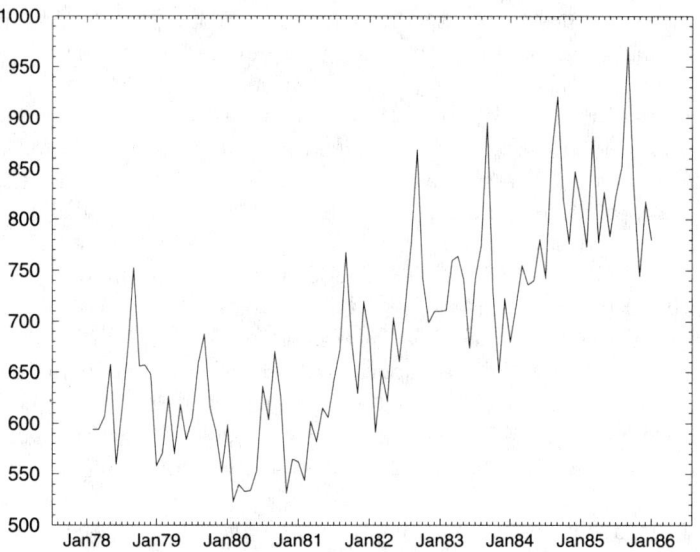

Figure 4.6. *Edendale births: monthly totals of deliveries, February 1978 – January 1986.*

with
$$\log {}_t\lambda_i = a_i + bt; \quad (4.8)$$
and one incorporating in addition sinusoidal terms at the annual frequency,
$$\log {}_t\lambda_i = a_i + bt + c\cos(2\pi t/12) + d\sin(2\pi t/12). \quad (4.9)$$

Both models fitted are unsuccessful in the sense that they effectively degenerate to one-state models: in each case one of the two off-diagonal transition probabilities is so close to zero as to be negligible, and the stationary distribution assigns probability one to one of the two states.

A number of logistic-linear models were then fitted by GLIM, and it is notable that GLIM yielded (*inter alia*) precisely the two models described above, with parameter estimates agreeing to four significant figures, but required fewer parameters to do so. (For instance, the general two-state hidden Markov model with conditional mean given by equation (4.9) has a total of seven parameters, and the corresponding GLIM model has four.) Table 4.12 compares the models fitted by GLIM to the total deliveries, and

Table 4.12. *Edendale births: models fitted by GLIM to the log of the mean number of deliveries.*

Explanatory variables	Deviance	BIC
t *	545.98	1358.1
t, sinusoidal terms **	428.70	1249.9
t, sinusoidal terms, n_{t-1}	356.10	**1181.9**
t, sinusoidal terms, n_{t-1}, n_{t-2}	353.91	1184.3
sinusoidal terms, n_{t-1}	464.83	1286.1
t, n_{t-1}	410.69	1227.3
n_{t-1}	499.47	1311.6

* This is the model identical to the hidden Markov model (4.8).
** This is identical to model (4.9).

from that table it can be seen that BIC selects the model incorporating time trend, sinusoidal components at the annual frequency, and the number of deliveries in the previous month (n_{t-1}). The details of this model are as follows. Conditional on the history, the number of deliveries in month t (N_t) is distributed Poisson with mean $_t\lambda$, where

$$\log {}_t\lambda = 6.015 + 0.002436t - 0.03652\cos(2\pi t/12)$$
$$- 0.02164\sin(2\pi t/12) + 0.0005737 n_{t-1}.$$

(Here, as before, February 1978 is month 1.)

Although both the hidden Markov models and the logistic-linear autoregressions reveal time trend and seasonality in this case, only the latter are able to detect dependence on the previous observation. This suggests that the dependence which is present in the total deliveries series is simply of a kind that can be detected by appropriate observation-driven models, but not by parameter-driven models.

4.4.3 Conclusion

The conclusion is therefore twofold. If a model for the number of Caesareans, given the total number of deliveries, is needed, the binomial-hidden Markov model with time trend is best of all of those considered (including various logistic-linear autoregressive models). If on the other hand a model for the total deliveries is needed (e.g. as a building-block in projecting the number of Caesareans) a first-order logistic-linear autoregressive model incorporating also time trend and sinusoidal components seems best, and is certainly superior to the Poisson-hidden Markov models fitted.

4.5 Locomotory behaviour of *Locusta migratoria*

In the study of animal behaviour sequences, discrete-time Markov chains of first order or higher are quite commonly used to model the successive behaviour categories of a wide variety of species: for instance blowflies (Cane, 1978), beavers (Rugg and Buech, 1990) and rhesus monkeys (Cane, 1978). General discussions of such modelling are provided by Chatfield and Lemon (1970), Chatfield (1973), Morgan (1976) and Cane (1978). The two main purposes of such modelling are firstly to provide a fairly simple summary of the behaviour sequence observed, and secondly to provide a basis for comparisons between subjects or between the behaviours of a single subject under different conditions.

It seems usually to be assumed in such applications that the transition probabilities are stationary, i.e. that a homogeneous Markov chain is appropriate as a model. Clearly this assumption would not be justified if there were a trend in environmental conditions or in the motivational state of the animal under observation. A further limitation of the usual approach is that it is not easily applicable to joint modelling of several subjects with possibly differing transition probabilities. (If it may reasonably be assumed that several independent subjects possess the same transition probabilities, these probabilities can then be estimated by pooling the transition counts across subjects. If on the other hand the transition probabilities may differ, such pooling may not be meaningful.) Expansion of the state space of the Markov chain to accommodate several subjects greatly increases its dimension and *a fortiori* the number of transition probabilities being estimated, and it is doubtful whether such a complex model would be useful in the role of summarizing behaviour. Hidden Markov models can, however, cope both with time trend and with expansion to several subjects without an explosion in the number of parameters. Furthermore, they provide an alternative to Markov chains of order greater than one if it is thought that the Markov property (of order one) is too restrictive an assumption and a longer memory is needed for a model to be realistic.

In this section we describe multivariate and univariate hidden Markov models fitted to the simultaneous behaviour of 24 locusts

(*Locusta migratoria*) in an experiment carried out by Dr D. Raubenheimer. The experiment investigated the effect of hunger on locomotory behaviour. The subjects were all three days into the fifth larval stadium, and the experiment was conducted as follows. The locusts were placed individually in glass observation chambers. The even-numbered subjects were deprived of food for $5\frac{1}{2}$ hours before observation commenced, during which period the odd-numbered subjects were allowed to feed *ad libitum*. Food and water were not available during observation. Within each of the two groups the subjects were alternately male and female. During the observation period data were entered directly into a computer, which had been programmed to emit a signal at 30-second intervals and thereafter to accept an entry for the observed behaviour of each of the subjects: locomoting, quiescent or (rarely) grooming. The number of observations made on each subject was 161. For the purpose of this analysis the categories used were locomoting (denoted by 1) and not locomoting (0), the latter category including both quiescent and grooming behaviour.

As a preliminary exercise, one-step transition counts were found for each of the 24 subjects, and homogeneous Markov chain models fitted thereby. Although these models do indeed point to differences between the fed and the unfed subjects (all but one of the unfed subjects having a higher unconditional probability of locomoting than all the fed subjects), there are two facts which suggest that a homogeneous Markov chain may not be an appropriate model. Firstly, in most of the 24 cases the sample ACF of the observed binary sequence is very far from being of the form α^k implied by a (stationary) two-state Markov chain. Secondly, the level of activity across all subjects increases with the passage of time, which suggests that there may be a time trend in the transition probabilities. At the very least, therefore, we can conclude that a *stationary* homogeneous Markov chain is not appropriate.

4.5.1 Multivariate models

It was therefore decided to fit a two-state multivariate hidden Markov model with time trend to each of the two groups of 12 subjects, and to compare these models with similar ones lacking a time trend, in order to assess whether inclusion of such a trend appreciably improves the fit. The model without trend consists of a two-state Markov chain and the 24 state-dependent probabilities of locomotion, one for each of the 12 subjects in each of the two states. Such a model has 26 parameters. For the fed subjects, the multivariate model without time trend has transition probability matrix $\begin{pmatrix} 0.995 & 0.005 \\ 0.009 & 0.991 \end{pmatrix}$, corresponding stationary distribution (0.627, 0.373), and log-likelihood (l) equal to -371.81. The 24 'state-dependent' probabilities are the probabilities of locomotion for each of the 12 subjects in each of the two states. The model fitted has the property that, for nine of the 12 subjects, state 1 has associated with it a lower probability of locomotion than has state 2 — but of course it must be remembered that the numbering of states is arbitrary.

The more general model, which allows for a single time trend (i.e. the same trend for all subjects in the two states), assumes that the probability of locomotion of subject j in state i at time t is $_tp_{ij}$, where for $i = 1, 2$ and $j = 1, 2, \ldots, 12$

$$\text{logit } _tp_{ij} = a_{ij} + bt.$$

Such a model has 27 parameters, and the model fitted has a log-likelihood of -365.40, whence by AIC and BIC it is preferable to the model without time trend. (In passing, we note that a model allowing b to vary between states was also fitted, but because it resulted in a very small improvement in the log-likelihood, to -365.20, it was not considered further.) The 27-parameter model fitted is as follows. The underlying Markov chain has transition probability matrix $\begin{pmatrix} 0.994 & 0.006 \\ 0.011 & 0.989 \end{pmatrix}$ and stationary distribution (0.654, 0.346). The time-trend parameter b is 0.01405, and the parameters a_{ij} are given for $i = 1, 2$ and $j = 1, 2, \ldots, 12$ in Table 4.13. There is no clear pattern of these parameters being greater for one state than another: for seven of the 12 subjects a_{1j} exceeds a_{2j}.

A similar hierarchy of three models was fitted to the behaviour of the 12 unfed subjects. The 26-parameter model had $l = -1116.1$, the 27-parameter $l = -1104.5$, and the 28-parameter $l = -1103.6$. As in the case of the fed subjects, therefore, we concentrate on the 27-parameter model. The Markov chain of that model has transition probability matrix $\begin{pmatrix} 0.966 & 0.034 \\ 0.064 & 0.936 \end{pmatrix}$ and stationary distribution (0.653, 0.347). The trend parameter b is 0.01054, and the parameters a_{ij} for this model are also given in Table 4.13. It will be noted that, for all but two of the unfed subjects, a_{1j} exceeds a_{2j} and $_tp_{1j}$ therefore exceeds $_tp_{2j}$ for all t.

We now compare the two sets of subjects on the basis of these

Table 4.13. *Locomotion of locusts: fed subjects (left) and unfed subjects (right). Parameters a_{ij} of multivariate hidden Markov models with single time trend in each case.*

j	a_{1j}	a_{2j}	j	a_{1j}	a_{2j}
1	−5.841	−23.47	1	−0.002	−2.292
2	−3.996	−1.139	2	−0.999	−1.656
3	−4.210	−4.797	3	1.803	−1.386
4	−4.193	−4.402	4	0.917	−0.205
5	−5.841	−4.086	5	−0.588	0.099
6	−5.841	−5.566	6	−2.092	−21.37
7	−3.066	−3.098	7	−0.311	−1.001
8	−4.722	−23.49	8	−1.332	−1.748
9	−25.41	−3.425	9	−1.229	−2.268
10	−3.559	−1.641	10	0.144	−0.394
11	−2.904	−2.966	11	−1.325	−2.325
12	−5.841	−23.48	12	−1.471	−1.324

two 27-parameter models. Because the same trend parameter applies whether the state is 1 or 2, it is meaningful to compare time trends between the two sets of subjects even though the underlying Markov chains differ. (As it happens, the stationary distributions of the two Markov chains are practically identical, but that is not necessary for a comparison to be meaningful.)

Although the trend is positive in both cases (i.e. the probability of locomotion increases with the passage of time), it is greater (on a logit scale) for the fed subjects than for the unfed ones. This is plausible: the subjects fed beforehand are experiencing a greater change in their condition during the observation period than are the unfed subjects. Furthermore, there appears to be among the unfed subjects a greater consistency of behaviour than is the case for the fed ones: the evidence for this statement is the property noted above that, for all but two of the unfed subjects, $_tp_{1j}$ exceeds $_tp_{2j}$. For the fed subjects the situation is less clear-cut, in that a_{1j} exceeds a_{2j} for seven of the 12 subjects. If indeed the behaviour of the unfed subjects is more consistent, this would not be surprising: the fed subjects have entered observation at varying stages of the 'cycle of satiety', whereas the others were standardized by $5\frac{1}{2}$ hours of food deprivation.

Table 4.14. *Locust data: comparison of univariate models pooling movements within groups.*

model	k	$-l$	AIC	BIC
fed subjects, no time trend	4	169.80	347.6	359.9
fed subjects, one trend	5	163.93	**337.9**	**353.3**
unfed subjects, no trend	4	314.07	636.1	648.5
unfed subjects, one trend	5	304.49	**619.0**	**634.4**

4.5.2 Univariate models

Although in the case of the data considered here there is separate information on each of the subjects, there are experimental situations in which it is known only, for each time, *how many* of the subjects have a given characteristic: that is, only 'macro data' are available. Furthermore, even if 'micro data' are available, it may be the case that interest resides not in individual characteristics of the subjects under observation but in what may be concluded of them as a group. It is therefore interesting to fit univariate models to the number locomoting in each group, to make comparisons between these univariate models for the two groups, and to compare the conclusions with those drawn from the multivariate models discussed above.

We consider, for each group of 12 subjects, univariate models which pool the movements of the subjects. First a simple two-state binomial-hidden Markov model was fitted to the series of length 161 giving for each time point the number of subjects (out of 12) observed to be locomoting. In the notation of Chapter 2, these models have n_t equal to 12 for all t, and they can of course be generalized by incorporating time trends in the state-dependent probabilities. For the fed and unfed subjects respectively, the models without trend achieved log-likelihood values of -169.80 and -314.07. The corresponding models with a single time trend applying to both states achieved values of -163.93 and -304.49. Models allowing for differing time trends in the two states were also fitted, but produced only very minor improvements in the log-likelihood (to -163.90 and -303.66) and were not considered further. The models with a single trend are clearly superior to those without time trend, on the basis of AIC or BIC: see Table 4.14.

In detail, the models with a single time trend are as follows.

For the fed subjects, the underlying Markov chain has transition probability matrix $\begin{pmatrix} 0.963 & 0.037 \\ 0.020 & 0.980 \end{pmatrix}$, stationary distribution (0.355, 0.645), and probabilities $_tp_i$ of locomotion at time t in state i specified by

$$\text{logit } _tp_1 = -4.745 + 0.01788t$$

and

$$\text{logit } _tp_2 = -4.050 + 0.01788t.$$

For the unfed subjects, the t.p.m. is $\begin{pmatrix} 0.986 & 0.014 \\ 0.013 & 0.987 \end{pmatrix}$, the stationary distribution is (0.478, 0.522), and

$$\text{logit } _tp_1 = -1.280 + 0.008747t$$

and

$$\text{logit } _tp_2 = -0.5103 + 0.008747t.$$

Again the trend is positive for both fed and unfed subjects, and greater (on a logit scale) for the fed subjects. The initial probability of locomotion is very much smaller for the fed subjects (0.009 or 0.017, depending on the state) than it is for the unfed ones (0.218 or 0.375). The corresponding probabilities for time 161 are 0.134 and 0.237 (fed) and 0.532 and 0.711 (unfed).

4.5.3 Conclusion

We may therefore conclude that both types of hidden Markov model (multivariate and univariate) can in this application provide a relatively simple and meaningful summary, and are useful as bases for comparison between groups. The great advantage that the hidden Markov models have over straightforward Markov chain models is that they can in fairly simple fashion accommodate several subjects and dependence on time (or any other covariate).

For the purposes of summary and comparison of behaviour, it is not essential that the hidden Markov models should possess a substantive interpretation. An interpretation of the multivariate model that may prove useful, however, is as follows. With each subject, each state of the Markov chain and each time-point, there is associated a probability distribution for the behaviour of that subject at that time. We may interpret this probability distribution as representing the motivational state of the subject at that time. Knowledge of that distribution and the current state of the Markov chain would not enable us to predict with certainty the

168   APPLICATIONS

behaviour of the subject at the relevant time, but it would reduce the uncertainty and tell us, for instance, whether locomotion was very likely at that time.

4.6 Wind direction at Koeberg

South Africa's only nuclear power station is situated at Koeberg on the west coast, about 30 km north of Cape Town. Wind direction, wind speed, rainfall and other meteorological data are collected continuously by the Koeberg weather station with a view to their use in radioactive plume modelling, *inter alia*. Four years of wind direction data were made available by the staff of the Koeberg weather station, and this section describes an attempt to model the wind direction at Koeberg by means of hidden Markov models for categorical time series.

The data consist of hourly values of the average wind direction over the preceding hour at 35 m above ground level. The period covered is 1 May 1985 to 30 April 1989 inclusive. The average referred to is a vector average, which allows for the circular nature of the data, and is given in degrees. There are in all 35 064 observations; there are no missing values. Before any models were fitted the hourly averages were classified into the 16 conventional directions N, NNE, ..., NNW, coded 1 to 16 in that order.

4.6.1 Three hidden Markov models for hourly averages of wind direction

The first model fitted was a simple multinomial-hidden Markov model with two states and no seasonal components, the case $m=2$ and $q=16$ of the categorical model described in section 3.3.3. In this model there are 32 parameters to be estimated: two transition probabilities to specify the Markov chain, and 15 probabilities for each of the two states, subject to the sum of the 15 not exceeding one. The results are as follows. The underlying Markov chain has transition probability matrix $\begin{pmatrix} 0.964 & 0.036 \\ 0.031 & 0.969 \end{pmatrix}$ and stationary distribution (0.462, 0.538), and the 16 probabilities associated with each of the two states are displayed in Table 4.15. A graph of these two sets of probabilities appears as Figure 4.7. The value of the unconditional log-likelihood achieved by this model is $-75\,832.1$.

The model successfully identifies two apparently meaningful weather states which are very different at least as regards the likely

WIND DIRECTION AT KOEBERG

Table 4.15. *Koeberg wind data (hourly): probabilities of each direction in the simple two-state hidden Markov model.*

1	N	0.129	0.000
2	NNE	0.048	0.000
3	NE	0.059	0.001
4	ENE	0.044	0.026
5	E	0.006	0.050
6	ESE	0.001	0.075
7	SE	0.000	0.177
8	SSE	0.000	0.313
9	S	0.001	0.181
10	SSW	0.004	0.122
11	SW	0.034	0.050
12	WSW	0.110	0.008
13	W	0.147	0.000
14	WNW	0.130	0.000
15	NW	0.137	0.000
16	NNW	0.149	0.000

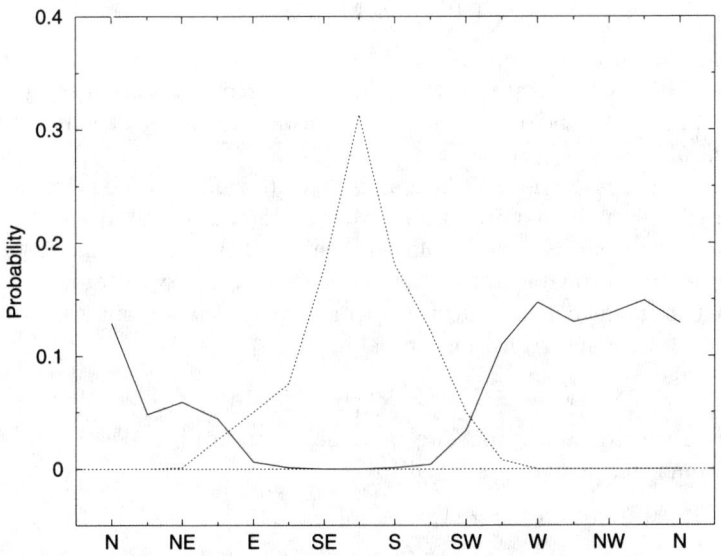

Figure 4.7. *Koeberg wind data (hourly): probabilities of each direction in the simple two-state hidden Markov model.*

wind directions in those states. In state 1 the most likely direction is NNW, and the probability falls away on either side of NNW, reaching a level of less than 0.01 for all directions (clockwise) from E to SSW inclusive. In state 2 the peak is at direction SSE, and the probability falls away sharply on either side, being less than 0.01 for all directions from WSW to NE inclusive. Two generalizations of this type of model were also fitted: firstly, a model based on a three-state Markov chain rather than a two-state one; and secondly, a model based on a two-state Markov chain but incorporating both a daily cycle and an annual cycle. We shall now describe these two models, and in due course compare all the models considered on the basis of AIC and BIC.

The three-state hidden Markov model has 51 parameters: six to specify the Markov chain, and 15 probabilities associated with each of the states. Essentially this model splits state 2 of the two-state model into two new states, one of which peaks at SSE and the other at SE. The transition probability matrix is

$$\begin{pmatrix} 0.957 & 0.030 & 0.013 \\ 0.015 & 0.923 & 0.062 \\ 0.051 & 0.077 & 0.872 \end{pmatrix}$$

and the stationary distribution is (0.400, 0.377, 0.223). The 16 probabilities associated with each of the three states are displayed in Table 4.16 and Figure 4.8. The unconditional log-likelihood is $-69\,525.9$.

As regards the model which adds daily and annual cyclical effects to the simple two-state hidden Markov model, it was decided to build these effects into the Markov chain rather than into the state-dependent probabilities. This was because, for a two-state chain, we can model cyclical effects parsimoniously by assuming that the two off-diagonal transition probabilities

$$P(C_t \neq i \mid C_{t-1} = i) = {}_t\gamma_i$$

are given by an appropriate periodic function of t. We therefore assume that logit ${}_t\gamma_i$ is equal to

$$a_i + b_i \cos(2\pi t/24) + c_i \sin(2\pi t/24) \\ + d_i \cos(2\pi t/8766) + e_i \sin(2\pi t/8766) \quad (4.10)$$

for $i = 1, 2$ and $t = 2, 3, \ldots, T$. A similar model for each of the state-dependent probabilities in each of the two states would involve many more parameters, or a problem of asymmetry which does not arise if the above method is used, or both. As discussed

WIND DIRECTION AT KOEBERG

Table 4.16. *Koeberg wind data (hourly): probabilities of each direction in the three-state hidden Markov model.*

1	N	0.148	0.000	0.001
2	NNE	0.047	0.000	0.016
3	NE	0.016	0.000	0.097
4	ENE	0.003	0.000	0.148
5	E	0.001	0.000	0.132
6	ESE	0.000	0.000	0.182
7	SE	0.000	0.023	0.388
8	SSE	0.000	0.426	0.033
9	S	0.000	0.257	0.002
10	SSW	0.002	0.176	0.000
11	SW	0.020	0.089	0.000
12	WSW	0.111	0.028	0.000
13	W	0.169	0.002	0.000
14	WNW	0.151	0.000	0.000
15	NW	0.159	0.000	0.001
16	NNW	0.173	0.000	0.000

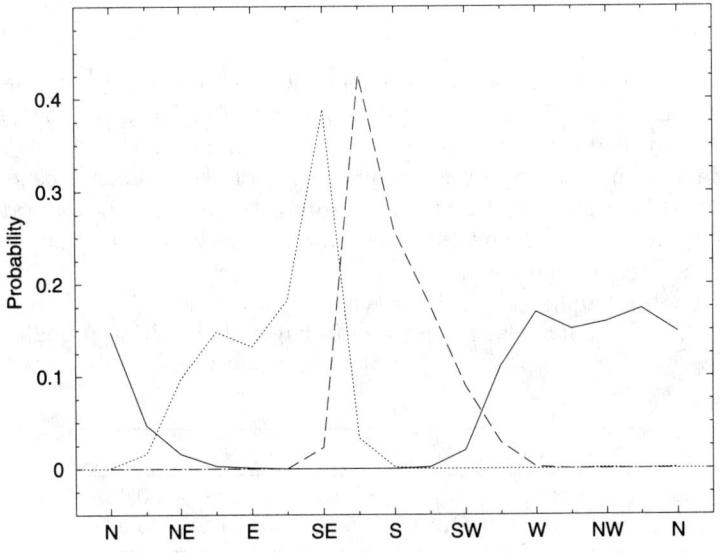

Figure 4.8. *Koeberg wind data (hourly): probabilities of each direction in the three-state hidden Markov model.*

Table 4.17. *Koeberg wind data (hourly): probabilities of each direction in the two-state hidden Markov model with cyclical components.*

1	N	0.127	0.000
2	NNE	0.047	0.000
3	NE	0.057	0.002
4	ENE	0.027	0.040
5	E	0.004	0.052
6	ESE	0.001	0.076
7	SE	0.001	0.179
8	SSE	0.000	0.317
9	S	0.001	0.183
10	SSW	0.007	0.121
11	SW	0.059	0.026
12	WSW	0.114	0.003
13	W	0.145	0.000
14	WNW	0.128	0.000
15	NW	0.135	0.000
16	NNW	0.147	0.000

in section 3.7, the estimation technique has to be modified when the underlying Markov chain is not assumed to be homogeneous. The estimates in this case were based on the initial state of the Markov chain being state 2, and the log of the likelihood conditioned on that initial state is $-75\,658.5$. (Conditioning on state 1 yielded a slightly inferior value for the likelihood, and similar parameter estimates.)

The probabilities associated with each state are given in Table 4.17, and the models for the two off-diagonal transition probabilities are given below, in the notation of equation (4.10).

i	a_i	b_i	c_i	d_i	e_i
1	-3.349	0.197	-0.695	-0.208	-0.401
2	-3.523	-0.272	0.801	0.082	-0.089

From Table 4.17 it will be noted that the general pattern of the state-dependent probabilities is very similar to the pattern seen in the simple two-state model without any cyclical components.

Table 4.18. *Koeberg wind data (hourly): transition probability matrix of saturated Markov chain model.*

	1	2	3	4	5	6	7	8	9	10	11	12	13	14	15	16
row 1	0.610	0.080	0.022	0.008	0.003	0.001	0.001	0.002	0.004	0.000	0.003	0.003	0.014	0.020	0.037	0.190
row 2	0.241	0.346	0.163	0.037	0.015	0.001	0.003	0.003	0.003	0.003	0.009	0.010	0.029	0.037	0.036	0.064
row 3	0.056	0.164	0.468	0.134	0.028	0.006	0.009	0.006	0.004	0.006	0.006	0.018	0.032	0.018	0.019	0.025
row 4	0.013	0.033	0.163	0.493	0.144	0.032	0.017	0.011	0.011	0.017	0.007	0.012	0.014	0.014	0.010	0.010
row 5	0.009	0.007	0.048	0.249	0.363	0.138	0.053	0.027	0.033	0.022	0.011	0.006	0.008	0.007	0.013	0.009
row 6	0.004	0.009	0.010	0.051	0.191	0.423	0.178	0.051	0.025	0.023	0.008	0.007	0.004	0.006	0.004	0.004
row 7	0.001	0.001	0.003	0.006	0.023	0.141	0.607	0.160	0.036	0.008	0.006	0.003	0.002	0.001	0.001	0.001
row 8	0.001	0.001	0.001	0.003	0.005	0.016	0.140	0.717	0.094	0.013	0.005	0.003	0.002	0.001	0.000	0.000
row 9	0.004	0.001	0.002	0.004	0.005	0.008	0.025	0.257	0.579	0.077	0.017	0.009	0.005	0.003	0.001	0.000
row 10	0.002	0.002	0.002	0.001	0.006	0.005	0.010	0.036	0.239	0.548	0.093	0.041	0.010	0.005	0.002	0.000
row 11	0.005	0.002	0.003	0.005	0.003	0.003	0.008	0.012	0.038	0.309	0.397	0.151	0.038	0.012	0.008	0.005
row 12	0.004	0.002	0.002	0.003	0.001	0.003	0.002	0.005	0.017	0.056	0.211	0.504	0.149	0.019	0.016	0.007
row 13	0.010	0.005	0.005	0.003	0.004	0.001	0.002	0.005	0.004	0.013	0.028	0.178	0.561	0.138	0.030	0.013
row 14	0.013	0.005	0.004	0.003	0.003	0.003	0.001	0.001	0.001	0.007	0.008	0.027	0.188	0.494	0.199	0.043
row 15	0.031	0.009	0.007	0.004	0.005	0.002	0.001	0.001	0.002	0.001	0.003	0.011	0.043	0.181	0.509	0.190
row 16	0.158	0.023	0.009	0.005	0.001	0.001	0.002	0.001	0.000	0.002	0.002	0.004	0.017	0.054	0.162	0.559

4.6.2 Model comparisons and other possible models

The three models described above were compared with each other and with a saturated 16-state Markov chain model, on the basis of AIC and BIC. The transition probabilities defining the Markov chain model were estimated by conditional maximum likelihood, as described in section 1.2.1, and are displayed in Table 4.18. The comparison appears as Table 4.19.

What is of course striking in Table 4.19 is that the likelihood of the saturated Markov chain model is so much higher than that

Table 4.19. *Koeberg wind data (hourly): comparison of four models fitted.*

model	k	$-l$	AIC	BIC
2-state HM	32	75 832.1	151 728	151 999
3-state HM	51	69 525.9	139 154	139 585
2-state HM with cycles	40	75 658.5*	151 397	151 736
saturated Markov chain	240	48 301.7	**97 083**	**99 115**

* conditional on state 2 being the initial state

of the hidden Markov models that the large number of parameters of the Markov chain (240) is virtually irrelevant when comparisons are made by AIC or BIC. It is therefore interesting to compare certain properties of the Markov chain model with the corresponding properties of the simple two-state hidden Markov model (e.g. unconditional and conditional probabilities). The unconditional probabilities of each direction were computed for the two models, and are almost identical.

However, examination of the conditional probability $P(S_t = 16 \mid S_{t-1} = 8)$, where S_t denotes the direction at time t, points to an important difference between the models. For the Markov chain model this probability is zero, since no such transitions were observed. For the hidden Markov model it is, in the notation of equation (3.8):

$$\frac{\delta P_{(8)} \Gamma P_{(16)} \mathbf{1}'}{\delta P_{(8)} \mathbf{1}'} = \frac{0.5375 \times 0.3131 \times 0.0310 \times 0.1494}{0.5375 \times 0.3131} = 0.0046.$$

Although small, this probability is not insignificant. The observed number of transitions from SSE (direction 8) was 5899. On the basis of the hidden Markov model one would therefore expect about 27 of these transitions to be to NNW. None were observed. Under the hidden Markov model 180-degree switches in direction are quite possible. Every time the state changes (which happens at any given time point with probability in excess of 0.03), the most likely direction of wind changes by 180 degrees. This is inconsistent with the observed gradual changes in direction: the matrix of observed transition counts is heavily dominated by diagonal and near-diagonal elements (and, because of the circular nature of the categories, elements in the corners furthest from the principal diagonal). Changes through 180 degrees were in general rarely observed.

Table 4.20. *Koeberg wind data (daily): comparison of two models fitted.*

model	k	$-l$	AIC	BIC
2-state HM	32	3461.88	**6987.75**	**7156.93**
saturated MC	240	3292.52	7065.04	8333.89

The above discussion suggests that, if daily figures are examined rather than hourly, the observed process will be more amenable to modelling by means of a hidden Markov model, because abrupt changes of direction are more likely in daily data than hourly. Markov chain and two-state hidden Markov models were therefore fitted to the series of length 1461 beginning with the first observation and including every 24th observation thereafter. For these data the hidden Markov model proved to be superior to the Markov chain even on the basis of AIC, which penalizes extra parameters less here than does BIC: see Table 4.20. Although a daily model is of little use in the main application intended (evacuation planning), there are other applications for which a daily model is exactly what one needs, e.g. forecasting the wind direction a day ahead.

Since the (first-order) Markov chain model does not allow for dependence beyond first order, the question that arises is whether any model for the hourly data which allows for higher-order dependence is superior to the Markov chain model: the hidden Markov models considered clearly are not. A saturated second-order Markov chain model would have an excessive number of parameters, and the Pegram model for (e.g.) a second-order Markov chain cannot reflect the property that, if a transition is made out of a given category, a nearby category is a more likely destination than a distant one.

The Raftery models do not suffer from that disadvantage, and in fact it is very easy to find a lag-2 Raftery model which is convincingly superior to the first-order Markov chain model. In the notation of section 1.3, take Q to be the transition probability matrix of the first-order Markov chain, as displayed in Table 4.18, and perform a simple line-search to find that value of λ_1 which maximizes the resulting conditional likelihood. This turns out to be 0.925. With these values for Q and λ_1 as starting values, the conditional likelihood was then maximized with respect to all 241 parameters, subject to the assumption that $0 \leq \lambda_1 \leq 1$. (This as-

Table 4.21. *Koeberg wind data (hourly): comparison of first-order Markov chain with lag-2 Raftery models.*

model	k	$-l$	AIC	BIC
saturated MC	240	48 301.7	97 083.4	99 115.0
Raftery model with starting values	241	48 087.8*	96 657.5	98 697.6
Raftery model fitted by max. likelihood	241	48 049.8*	**96 581.6**	**98 621.7**

* conditioned on the first two states

sumption makes it unnecessary to impose 16^3 nonlinear constraints on the maximization, and seems reasonable in the context of hourly wind directions showing a high degree of persistence.) The resulting value for λ_1 is 0.9125, and the resulting matrix Q does not differ much from its starting value. The log-likelihood achieved is $-48\,049.8$: see Table 4.21 for a comparison of likelihood values and the usual model selection criteria, from which the Raftery model emerges as superior to the Markov chain.

4.6.3 Conclusion

The general conclusion we may draw from the above analysis is that, both for the hourly and the daily wind direction data, it is possible to fit a model which is superior (in terms of the model selection criteria used) to a saturated first-order Markov chain. In the case of the hourly data, a lag-2 Raftery model (i.e. a particular kind of second-order Markov chain model) is preferred. In the case of the daily data, a simple two-state hidden Markov model for categorical time series, as introduced in section 3.3.3, performs better than the Markov chain.

One further approach to the hourly data that might well turn out to be superior to any of the models considered above is to develop a parsimonious class of models for Markov chains with transition probability matrix of the form seen in Table 4.18. This could be extended to higher-order Markov chains in exactly the same way as Raftery's models generalize the saturated (first-order) Markov chain.

4.7 Evapotranspiration

Kedem (1976) has proposed a second-order Markov chain model for a binary time series derived from a set of evapotranspiration data. The data are $\{Z_t : t = 1, 2, \ldots, 96\}$, rates of evapotranspiration in 96 consecutive months at a site in Israel, and are given in full by Kedem. (Evapotranspiration is the return of water vapour to the atmosphere by evaporation from land and water surfaces and by the transpiration of vegetation.) Kedem defines $W_t = Z_t - Z_{t-12}$ and

$$X_{t-12} = \begin{cases} 1 & \text{if } W_t \geq \overline{W} \\ 0 & \text{if } W_t < \overline{W}. \end{cases}$$

The resulting series $\{X_t\}$ has 84 values, and is given by:

11111 11100 01111 10000 00011 11000 10000 01111
    11110 01100 00000 11111 10000 00000 11111 11000 0000 .

Kedem performs a likelihood ratio test of the hypothesis of first order against second, and concludes that a second-order Markov chain is an appropriate model. He does, however, note that the significance level used is high (0.2). In order to discover whether a better model than a second-order Markov chain could be found, hidden Markov models (with two, three and four states) were therefore fitted to the series, as well as the model which assumes independence.

All the models (independence, first- and second-order Markov chains and the three hidden Markov models) were then compared by BIC, which is a consistent estimator of Markov chain order (Katz, 1981). The results are as in Table 4.22, with l denoting the log of the unconditional likelihood. (We have omitted from this comparison of models one further possibility: the Pegram–Raftery second-order Markov chain. As such a model is nested in the general second-order Markov chain, it cannot achieve a value of $-l$ better than 38.203. This corresponds to a BIC value of 89.70, which is higher than 88.73, the BIC value of the first-order Markov chain. The Pegram–Raftery model cannot therefore emerge as the 'best'.)

Since the Markov chain and second-order Markov chain were estimated by conditional maximum likelihood, and the other models by unconditional, comparison on the basis of the unconditional likelihood is slightly unfair to the first two models. What is interesting is that the simplest model other than independence, the first-order Markov chain, emerges as the best — contrary to Kedem's con-

Table 4.22. *Evapotranspiration data: comparison of models.*

model	k	−l	BIC
independence	1	58.201	120.83
Markov chain	2	39.935	**88.73**
second-order MC	4	38.203	94.13
2-state hidden Markov	4	39.905	97.53
3-state HM	9	37.845	115.57
4-state HM	16	35.187	141.27

clusion that a second-order Markov chain is needed. Furthermore, the two-state binomial-hidden Markov model fitted is simply the Markov chain with transition probability matrix $\begin{pmatrix} 0.823 & 0.177 \\ 0.185 & 0.815 \end{pmatrix}$. (Although this is not identical to the Markov chain model yielding the value $l = -39.935$, i.e. the Markov chain with transition probability matrix $\begin{pmatrix} 35/42 & 7/42 \\ 8/41 & 33/41 \end{pmatrix}$, the difference is explained by the use of conditional maximum likelihood in one case and unconditional in the other.) The fact that the best two-state hidden Markov model that can be found is just a Markov chain strengthens the conclusion that nothing more complicated than a Markov chain is justified, although the use of unconditional maximum likelihood seems preferable because the assumption of stationarity is reasonable for these data.

4.8 Thinly traded shares on the Johannesburg Stock Exchange

One of the difficulties encountered in modelling the price series of shares listed on the Johannesburg Stock Exchange is that many of the shares are only thinly traded. The market is heavily dominated by institutional investors, and if for any reason a share happens not to be an 'institutional favourite' there will very likely be days, or even weeks, during which no trading of that share takes place. One approach is to model the presence or absence of trading quite separately from the modelling of the price achieved when trading does take place. This is analogous to the modelling of the sequence of wet and dry days separately from the modelling of the amounts of precipitation occurring on the wet days. It is therefore natural to consider, as models for the trading pattern of one or several shares, hidden Markov models of the kind discussed by Zucchini and Guttorp (1991), who used them to represent the presence or absence of precipitation on successive days, at one or several sites.

In order to assess whether such models can be used successfully to represent trading patterns, data for six thinly traded shares were obtained from Dr D.C. Bowie, and various models, including two-state hidden Markov models, were fitted and compared. (The data appear as Table B.5 in Appendix B.) Of the six shares, three are from the coal sector and three from the diamonds sector. The coal shares are Amcoal, Vierfontein and Wankie, and the diamond shares Anamint, Broadacres and Carrigs. For all six shares the data cover the period from 5 October 1987 to 3 June 1991 (inclusive), during which time there were 910 days on which trading could take place. The data are therefore a multivariate binary time series of length 910.

4.8.1 Univariate models

The first two univariate models fitted to each of the six shares were a model assuming independence of successive observations and a Markov chain (the latter fitted by conditional maximum likelihood). In all six cases, however, the sample ACF bore little resemblance to the ACF of the Markov chain model, and the Markov chain was therefore considered unsatisfactory. Second-order Markov chains and two-state binomial-hidden Markov models, with and without trend, were also fitted, the former by conditional maximum likelihood and the latter by unconditional. The resulting log-likelihood and BIC values are shown in Table 4.23.

From that table we see that, of the five univariate models considered, the two-state hidden Markov model with a time trend fares best for four of the six shares: Amcoal, Vierfontein, Anamint and Broadacres. Of these four shares, all but Anamint show a negative trend in the probability of trading taking place, and Anamint a positive trend. In the case of Wankie, the model assuming independence of successive observations is chosen by BIC, and in the case of Carrigs a hidden Markov model without time trend is chosen.

Since a stationary hidden Markov model is chosen for Carrigs, it is interesting to compare the ACF of that model with the sample ACF and with the ACF of the competing Markov chain model. For the hidden Markov model the ACF is $\rho_k = 0.3517 \times 0.9127^k$,

Table 4.23. *Minus log-likelihood values and BIC values achieved by five types of univariate model for six thinly traded shares.*

Values of $-l$

model	Amcoal	Vierf'n	Wankie	Anamint	Broadac	Carrigs
independence	543.51	629.04	385.53	612.03	599.81	626.88
Markov chain	540.89	611.07	384.57	582.64	585.76	570.25
second-order M. chain	539.89	606.86	383.87	576.99	580.06	555.67
2-state HM, no trend	533.38	588.08	382.51	572.55	562.96	533.89
2-state HM with trend	528.07	577.51	381.28	562.55	556.21	533.88

Values of BIC

model	Amcoal	Vierf'n	Wankie	Anamint	Broadac	Carrigs
independence	1093.83	1264.89	**777.88**	1230.88	1206.43	1260.58
Markov chain	1095.41	1235.77	782.77	1178.91	1185.15	1154.13
second-order M. chain	1107.03	1240.97	794.99	1181.23	1187.37	1138.59
2-state HM, no trend	1094.01	1203.41	792.27	1172.35	1153.17	**1095.03**
2-state HM with trend	**1090.22**	**1189.08**	796.63	**1159.17**	**1146.49**	1101.83

Table 4.24. *Trading of Carrigs Diamonds: first eight terms of the sample ACF, compared with the autocorrelations of two possible models.*

ACF of Markov chain	0.350	0.122	0.043	0.015	0.005	0.002	0.001	0.000
sample ACF	0.349	0.271	0.281	0.237	0.230	0.202	0.177	0.200
ACF of HM model	0.321	0.293	0.267	0.244	0.223	0.203	0.186	0.169

and for the Markov chain it is $\rho_k = 0.3499^k$. Table 4.24 displays the first eight terms in each case. It is clear that the hidden Markov model comes much closer to matching the sample ACF than does the Markov chain model: a two-state hidden Markov model can model slow decay in ρ_k from any starting value ρ_1, but a two-state Markov chain cannot.

4.8.2 Multivariate models

Two-state multivariate hidden Markov models of two kinds were then fitted to each of the two groups of three shares: a model without time trend, and one which has a single (logistic-linear) time trend common to the two states and to the three shares in the

Table 4.25. *Comparison of several multivariate models for the three coal shares and the three diamond shares.*

Coal shares

model	k	$-l$	BIC
3 'independence' models	3	1558.08	3136.60
3 univariate HM models, no trend	12	1503.97	3089.69
3 univariate HM models with trend	15	1486.86	**3075.93**
multivariate HM model, no trend	8	1554.01	3162.52
multivariate HM, single trend	9	1538.14	3137.60

Diamond shares

model	k	$-l$	BIC
3 'independence' models	3	1838.72	3697.88
3 univariate HM models, no trend	12	1669.40	3420.56
3 univariate HM models with trend	15	1652.64	3407.48
multivariate HM model, no trend	8	1590.63	3235.77
multivariate HM, single trend	9	1543.95	**3149.22**

group. The first type of model has eight parameters, the second has nine. These models were then compared with each other and with the 'product models' obtained by combining independent univariate models for the individual shares. The three types of product model considered were those based on independence of successive observations and those obtained by using the univariate hidden Markov models with and without trend. The results are displayed in Table 4.25.

It is clear that, for the coal shares, the multivariate modelling has not been a success: the model consisting of three independent univariate hidden Markov models with trend is 'best'. We therefore give these three univariate models in Table 4.26. In each of these models $_tp_i$ is the probability that the relevant share is traded on day t if the state of the underlying Markov chain is i, and logit $_tp_i = a_i + bt$.

For the diamond shares, the best model of those considered is the multivariate hidden Markov model with trend. In this model logit $_tp_{ij} = a_{ij} + bt$, where $_tp_{ij}$ is the probability that share j is traded on day t if the state is i. The transition probability matrix is $\begin{pmatrix} 0.998 & 0.002 \\ 0.001 & 0.999 \end{pmatrix}$, the trend parameter b is -0.003160, and the other parameters a_{ij} are as follows:

share	a_{1j}	a_{2j}
Anamint	1.756	1.647
Broadacres	0.364	0.694
Carrigs	1.920	-0.965

Table 4.26. *Univariate hidden Markov models (with trend) for the three coal shares.*

share	t.p.m.	a_1	a_2	b
Amcoal	$\begin{pmatrix} 0.774 & 0.226 \\ 0.019 & 0.981 \end{pmatrix}$	-0.332	1.826	-0.001488
Vierfontein	$\begin{pmatrix} 0.980 & 0.020 \\ 0.091 & 0.909 \end{pmatrix}$	0.606	3.358	-0.001792
Wankie	$\begin{pmatrix} 0.807 & 0.193 \\ 0.096 & 0.904 \end{pmatrix}$	-5.028	-0.943	-0.000681

4.8.3 Discussion

The parametric bootstrap, with a sample size of 100, was used to investigate the distribution of the estimators in the models displayed in Table 4.26. In this case the estimators show much more variability than do the estimators used for the geyser data analysed in section 4.2. Table 4.27 gives for each of the three coal shares the bootstrap sample means, medians and standard deviations for the estimators of the five parameters. It will be noted that the estimators of a_1 and a_2 seem particularly variable. It is, however, true that, except in the middle of the range, very large differences on a logit scale correspond to small ones on a probability scale: two models with very different values of a_1 (for instance) may therefore produce almost identical distributions for the observations. For all three shares the trend parameter b seems to be more reliably estimated than the other parameters. If one is interested in particular

Table 4.27. *Coal shares: means, medians and standard deviations of bootstrap sample of estimators of parameters of two-state hidden Markov models with time trend.*

share		$\hat{\gamma}_1$	$\hat{\gamma}_2$	$\hat{a}_1$	$\hat{a}_2$	$\hat{b}$
Amcoal	mean	0.251	0.048	−1.40	2.61	−0.00180
	median	0.228	0.024	−0.20	1.95	−0.00163
	s.d.	0.154	0.063	4.95	3.18	0.00108
Vierf'n	mean	0.023	0.102	0.599	4.06	−0.00187
	median	0.020	0.097	0.624	3.39	−0.00190
	s.d.	0.015	0.046	0.204	3.70	0.00041
Wankie	mean	0.145	0.148	−14.3	0.09	−0.00081
	median	0.141	0.089	−20.9	−0.87	−0.00074
	s.d.	0.085	0.167	10.3	4.17	0.00070

in whether trading is becoming more or less frequent, this will be the parameter of most substantive interest.

As regards the multivariate hidden Markov model for the three diamond shares, it is perhaps surprising that the model is so much improved by the inclusion of a single (negative) trend parameter: in the corresponding univariate models the time trend was positive for one share, negative and of similar magnitude for another share, and negligible for the remaining share. Another criticism to which this multivariate model is open is that the off-diagonal elements of the transition probability matrix of its underlying Markov chain are so close to zero as to be almost negligible: on average only one or two changes of state would take place during a sequence of 910 observations. Furthermore, it is not possible to interpret the two states as conditions in which trading is in general more likely and conditions in which it is in general less so. This is because the probability of trading $(_tp_{ij})$ is not consistently higher for one state i than the other.

In view of the relative lack of success of multivariate hidden Markov models in this application, these models are not pursued here. The above discussion does, we hope, serve as an illustration of the methodology, and suggests that such multivariate models are potentially useful in studies of this sort. They could, for instance, be used to model occurrences other than the presence or absence of trading, e.g. the price (or volume) rising above a given level.

4.9 Daily rainfall at Durban

The importance of understanding the stochastic structure of rainfall, or more generally of precipitation, hardly needs to be emphasized. In many countries rainfall is the element of climate most critical in its effects on land use and economic development. Precipitation data are routinely collected throughout the world, most commonly in the form of daily precipitation totals. Such data are used for the compilation of useful summaries, such as annual and monthly averages. However, in a wide variety of applications there are questions of interest (see for example Zucchini, Adamson and McNeill, 1992) that are difficult to answer unless one has available a model for the precipitation process.

Consider, for example, the question 'What is the probability that there is more than 200 mm rainfall in January and that the longest dry run is no longer than 5 days?'. Once a model is available, it is not difficult to estimate probabilities of this type, or other quantities of interest, by using Monte Carlo methods. One generates a very long realization from the model and then estimates the probability of the event of interest by its relative frequency in the generated rainfall sequence.

A review of the substantial literature on the modelling of daily precipitation is given by Woolhiser (1992). The standard method of constructing such models is to use two components, the first describing the occurrence or nonoccurrence of rainfall and the second the amounts of rainfall on wet days. In this section we restrict our attention to the first component. In other words, we consider models for the binary sequence of dry and wet days. Since such sequences exhibit marked seasonal behaviour as well as short-term persistence, we require models for seasonal binary time series.

We will consider models for the occurrence of rainfall at the Botanical Gardens in Durban, South Africa. As practically all precipitation at this site occurs in the form of rainfall, the terms 'rainfall' and 'precipitation', as used here, are synonymous. The series comprises daily rainfall totals for the period 1 January 1871 to 31 December 1992. There are, however, missing values amounting to about 11 years. Leap years are inconvenient when modelling rainfall series, and are usually truncated to 365 days. In what follows the observations for 29 February were removed — any rainfall that occurred on that day was added to the figure for 1 March.

A preliminary analysis of this series reveals, as one would expect, that the occurrence of rainfall exhibits seasonal behaviour. The probability of rain is greater in summer than it is in winter. Secondly, the series exhibits short-term persistence — the conditional probability that a wet day will be followed by a wet day is higher than the conditional probability that a dry day will be followed by a wet day. A simple model that captures both of these features is a two-state (nonhomogeneous) Markov chain whose transition probabilities vary cyclically over the year. There seems to be no evidence of long-term trend in this series.

Let S_t, $t = 1, 2, \ldots, T$ $(= 44\,530)$, be the binary random process representing rainfall occurrence, and let s_t, $t = 1, 2, \ldots, T$, be the corresponding observed values, where $s_t = 1$ if it rained on day t and $s_t = 0$ if it did not. To specify a Markov chain model for S_t one needs to specify the transition probability matrices:

$$\begin{pmatrix} \pi_{D|D}(t) & \pi_{W|D}(t) \\ \pi_{D|W}(t) & \pi_{W|W}(t) \end{pmatrix}$$
$$= \begin{pmatrix} P(S_t = 0 | S_{t-1} = 0) & P(S_t = 1 | S_{t-1} = 0) \\ P(S_t = 0 | S_{t-1} = 1) & P(S_t = 1 | S_{t-1} = 1) \end{pmatrix}$$

for $t = 1, 2, \ldots, T$. Since the row sums of these matrices are 1, we need specify only $\pi_{D|W}(t)$ and $\pi_{W|D}(t)$. Furthermore, on the assumption that the seasonal behaviour of the series remains unchanged over the years, it follows that $\pi_{D|W}(t) = \pi_{D|W}(t-365)$ for $t > 365$, and similarly $\pi_{W|D}(t)$. It is therefore sufficient to specify these functions for $t = 1, 2, \ldots, 365$, where day 1 is 1 January and day 0 is identified with day 365 (31 December).

An unconditional analysis requires one to specify also the initial probability distribution

$$(P(S_1 = 0), P(S_1 = 1)),$$

which, under certain conditions that we will not discuss here, is an awkward function of the sequence of transition probability matrices. In the discussion on hidden Markov models later in this section we will derive a simple approximation for this type of initial distribution, but for the moment we will circumvent the problem by conditioning the model on the first observation, that is we will regard $S_1 = s_1$ as given. An additional consideration is that the conditional models can be fitted by using standard software.

The transition probabilities $\pi_{W|D}(t)$ and $\pi_{D|W}(t)$ (for $t = 1, 2, \ldots, 365$) can be estimated by the corresponding relative frequencies:

$$\hat{\pi}_{D|W}(t) = N_{WD}(t)/N_{WO}(t), \text{ and}$$

$$\hat{\pi}_{W|D}(t) = N_{DW}(t)/N_{DO}(t),$$

where:

$N_{WD}(t)$ is the number of times day $t-1$ was wet and day t dry;

$N_{WO}(t)$ is the number of times day $t-1$ was wet and there was an observation on day t;

$N_{DW}(t)$ is the number of times day $t-1$ was dry and day t wet; and

$N_{DO}(t)$ is the number of times day $t-1$ was dry and there was an observation on day t.

These counts are given in Table B.6 of Appendix B.

The relative frequencies $\hat{\pi}_{D|W}(t)$ and $\hat{\pi}_{W|D}(t)$ are plotted as dots in Figure 4.9. These estimates are far from smooth — for example for days 220 and 221 one obtains $\hat{\pi}_{D|W}(220) = 0.765$ and $\hat{\pi}_{D|W}(221) = 0.333$. Abrupt changes of this magnitude reflect the instability of the estimators rather than the progress of seasonal fluctuation. The reason for this erratic behaviour of the estimates is that the model has an excessive number of parameters, namely $2 \times 365 = 730$.

One way to reduce the number of parameters and to obtain estimates of $\pi_{D|W}(t)$ and $\pi_{W|D}(t)$ that are smooth is to approximate each of the vectors (represented for convenience here by $\pi(t)$) by a truncated form of its Fourier series representation. Furthermore, to ensure that the estimate of $\pi(t)$ remains in the interval (0,1) it is convenient to work with its logit transformation:

$$\eta(t) = \operatorname{logit} \pi(t) = \log\left(\frac{\pi(t)}{1-\pi(t)}\right), \quad t = 1, 2, \ldots, 365.$$

Once $\eta(t)$ has been estimated, the probabilities can be computed by means of the inverse logit transformation:

$$\pi(t) = \frac{\exp(\eta(t))}{1+\exp(\eta(t))}, \quad t = 1, 2, \ldots, 365.$$

The following approximation reduces the number of parameters required to specify $\eta(t)$, and thus also $\pi(t)$, from 365 to $2q+1$:

$$\begin{aligned}
\eta(t) &= \alpha_0 + \sum_{i=1}^{187}\left(\alpha_i \cos\frac{2\pi i(t-1)}{365} + \beta_i \sin\frac{2\pi i(t-1)}{365}\right) \\
&\approx \alpha_0 + \sum_{i=1}^{q}\left(\alpha_i \cos\frac{2\pi i(t-1)}{365} + \beta_i \sin\frac{2\pi i(t-1)}{365}\right).
\end{aligned}$$

This is a generalized linear model with binomial error and logit

DAILY RAINFALL AT DURBAN

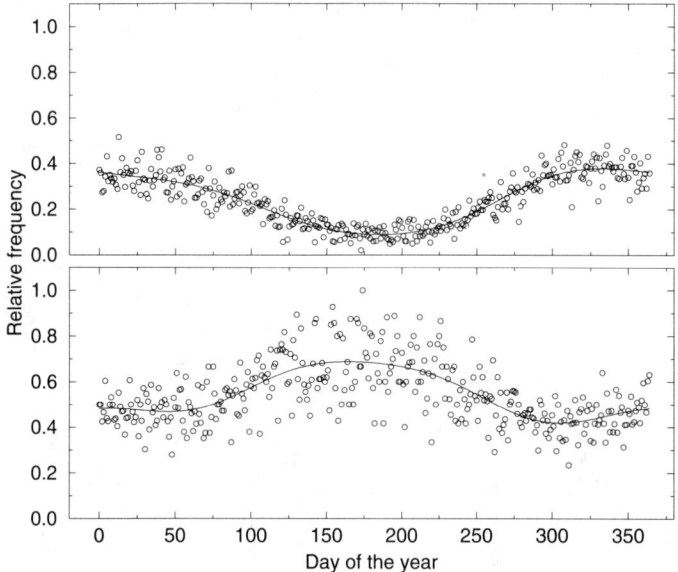

Figure 4.9. *Estimates of $\pi_{W|D}(t)$ (top) and $\pi_{D|W}(t)$ (bottom), $t = 1, 2, ..., 365$, using relative frequencies (circles) and the selected models (smooth curves).*

link (see for example McCullagh and Nelder, 1989) and so the parameters α_0, α_i, β_i ($i = 1, 2, \ldots, q$), can be estimated by using one of the available statistical software packages such as S-PLUS, which was used for the computations in this section. One generates, for $t = 1, 2, \ldots, 365$, values of the $2q$ explanatory variables:

$$x_{2i-1} = \cos\frac{2\pi i(t-1)}{365}, x_{2i} = \sin\frac{2\pi i(t-1)}{365}, \ i = 1, 2, \ldots, q.$$

The parameter α_i is the coefficient of x_{2i-1} in the model, β_i the coefficient of x_{2i}, and α_0 the intercept. To estimate these parameters for $\pi_{D|W}(t)$, the counts $N_{WD}(t)$ are taken to be the numbers of 'successes' in $N_{WO}(t)$ trials, $t = 1, 2, \ldots, 365$. For the case of $\pi_{W|D}(t)$ the counts $N_{DW}(t)$ and $N_{DO}(t)$ are applicable. An alternative method of parameter estimation is by direct maximization of the likelihood: see for example Zucchini and Adamson (1984).

The number of sine–cosine pairs (q) to be used in the above approximations needs to be specified. By increasing q one reduces

Table 4.28. *Values of minus log-likelihood based on* $\pi_{D|W}(t)$ *with* $q = 1, 2, 3, 4$ *(columns) and* $\pi_{W|D}(t)$ *with* $q = 1, 2, 3, 4$ *(rows)*.

q	1	2	3	4
1	22 491.78	22 442.39	22 441.44	22 441.26
2	22 468.42	22 419.04	22 418.08	22 417.90
3	22 464.86˙	22 415.47	22 414.52	22 414.34
4	22 464.60	22 415.21	22 414.26	22 414.08

Table 4.29. *AIC values corresponding to the models in Table 4.28.*

q	1	2	3	4
1	44 995.56	44 900.79	44 902.88	44 906.52
2	44 952.85	44 858.07	44 860.16	44 863.81
3	44 949.72	**44 854.95**	44 857.04	44 860.68
4	44 953.20	44 858.42	44 860.52	44 864.16

the approximation error, but then the standard error of the estimators increases. Ideally one should select q so as to minimize the combined effect of these two sources of discrepancy. A detailed account of this model selection problem is given in Chapter 10 of Linhart and Zucchini (1986). In what follows Akaike's Information Criterion (AIC) will be used to select the value of q.

Table 4.28 gives the value of minus log-likelihood for each of the 16 models obtained with $q = 1, 2, 3, 4$ for $\pi_{D|W}(t)$ and for $\pi_{W|D}(t)$. Table 4.29 gives the corresponding values of AIC.

The AIC value achieves a minimum when $q = 3$ for $\pi_{D|W}(t)$ and $q = 2$ for $\pi_{W|D}(t)$. The simpler model having $q = 2$ in both components minimizes BIC but, in effect, differs only slightly from that selected here, based on AIC. The parameter estimates for the selected model are displayed in Table 4.30. The resulting estimated conditional probabilities, $\hat{\pi}_{D|W}(t)$ and $\hat{\pi}_{W|D}(t)$, appear as smooth curves in Figure 4.9.

What we have described above is a two-state seasonal Markov chain model for a binary time series. The following two-state hidden Markov generalization of this model was also fitted to the series. The (hidden) Markov chain component of the model, $\{C_t : t = 1, 2, \ldots, T\}$, is characterized by the sequence of one-step transition

Table 4.30. *Parameter estimates for the selected model.*

	$\hat{\alpha}_0$	$\hat{\alpha}_1$	$\hat{\beta}_1$	$\hat{\alpha}_2$	$\hat{\beta}_2$	$\hat{\alpha}_3$	$\hat{\beta}_3$
$\hat{\eta}_{D\|W}(t)$	0.1861	−0.4885	0.1553	0.1616	−0.0182	0.0729	−0.0069
$\hat{\eta}_{W\|D}(t)$	−1.2485	0.8791	−0.0632	−0.1982	−0.0667		

probability matrices:

$$_t\Gamma = \begin{pmatrix} 1 - {_t\gamma_1} & {_t\gamma_1} \\ {_t\gamma_2} & 1 - {_t\gamma_2} \end{pmatrix}$$
$$= \begin{pmatrix} P(C_t = 1|C_{t-1} = 1) & P(C_t = 2|C_{t-1} = 1) \\ P(C_t = 1|C_{t-1} = 2) & P(C_t = 2|C_{t-1} = 2) \end{pmatrix},$$

and the state-dependent probabilities (for all t) by the two probabilities

$$\pi_i = P(S_t=1 \mid C_t=i), \quad i=1,2.$$

The seasonal fluctuation in the series can be modelled by approximating the logit transformations of each of $_t\gamma_1$ and $_t\gamma_2$ by its truncated Fourier representation. Omitting the subscripts 1, 2, one has:

$$_t\gamma = \frac{\exp(\nu(t))}{1 + \exp(\nu(t))},$$

where for $t=1, 2, \ldots, 365$:

$$\nu(t) \approx \alpha_0 + \sum_{i=1}^{q} \left(\alpha_i \cos \frac{2\pi i(t-1)}{365} + \beta_i \sin \frac{2\pi i(t-1)}{365} \right).$$

This seasonal hidden Markov chain model is not in the class of generalized linear models, for which estimation software is widely available. However, the parameters of the model, namely α_0, α_i, β_i ($i=1, 2, \ldots, q$), and π_1, π_2, can be estimated by direct maximization of the likelihood:

$$\begin{aligned} L &= P(S_1 = s_1, S_2 = s_2, S_3 = s_3, \ldots, S_T = s_T) \\ &= \delta \, \lambda(s_1) \, _2\Gamma \, \lambda(s_2) \, _3\Gamma \, \lambda(s_3) \cdots \, _T\Gamma \, \lambda(s_T) \mathbf{1}', \end{aligned}$$

where

$$\delta = (P(C_1 = 1), P(C_1 = 2))$$

and, for $t = 1, 2, ..., T$:

$$\lambda(s_t) = \begin{pmatrix} P(S_t = s_t | C_t = 1) & 0 \\ 0 & P(S_t = s_t | C_t = 2) \end{pmatrix}.$$

An exact expression for δ is difficult to obtain. However, if one makes the approximation $P(C_1 = 1) \approx P(C_0 = 1)$, it follows that:

$P(C_1 = 1)$
$= P(C_1 = 1 \mid C_0 = 1) P(C_0 = 1) + P(C_1 = 1 \mid C_0 = 2) P(C_0 = 2)$
$= (1 - {}_1\gamma_1) P(C_0 = 1) + {}_1\gamma_2 (1 - P(C_0 = 1))$
$\approx (1 - {}_1\gamma_1) P(C_1 = 1) + {}_1\gamma_2 (1 - P(C_1 = 1)).$

This, together with the analogous approximation for $P(C_1 = 2)$ — or, equivalently, by using $P(C_1 = 2) = 1 - P(C_1 = 1)$ — yields the conclusion that:

$$\delta \approx \frac{1}{{}_1\gamma_1 + {}_1\gamma_2} ({}_1\gamma_2, {}_1\gamma_1).$$

(This is just the stationary distribution of the homogeneous Markov chain with transition probability matrix ${}_1\Gamma$.) As is usual for a hidden Markov model, the expression for the likelihood remains valid if some of the values are missing. One simply replaces each λ matrix corresponding to a missing value by the 2×2 identity matrix.

The two-state Markov chain model described earlier in this section is the special case of the two-state hidden Markov model having $\pi_1 = 0$ and $\pi_2 = 1$. For the Durban rainfall series, the maximum likelihood estimate of π_1 is very close to zero and that of π_2 very close to one: in this case the hidden Markov generalization of the Markov chain model is therefore unwarranted.

Despite its lack of success in this application, the hidden Markov model outlined in this section is, we believe, a promising model in that it can be modified to describe other discrete-valued series exhibiting seasonal behaviour. The above formulas can easily be changed to accommodate monthly or quarterly series instead of daily series. Furthermore, by suitably modifying the state-dependent probabilities, several of the types of extension and modification discussed in Chapter 3 are possible. For example, one can model seasonal series of Poisson counts, multinomial or other multivariate series. The Markov chain component of the model, embodied in the sequence of matrices ${}_t\Gamma$ ($t = 1, 2, \ldots, 365$), takes care of the seasonal fluctuations.

Zucchini and Guttorp (1991) describe a hidden Markov model

for daily rainfall occurrence at several sites. In that paper seasonality is dealt with by partitioning the year into a small number of approximately homogeneous seasons. The parameters are taken to be constant within each season. An alternative model for such data, based on the model described here, would be to allow the sequence of matrices $_t\Gamma$ to vary smoothly and cyclically over the year.

4.10 Homicides and suicides, Cape Town, 1986–1991

In South Africa, as in the USA, gun control is a subject of much public interest and debate. Furthermore, there is in South Africa an apparently increasing tendency for violent crime to involve firearms. In a project intended to study this and related issues, Dr L.B. Lerer collected data relating to homicides and suicides from the South African Police mortuary in Salt River, Cape Town. Records relating to most of the homicide and suicide cases occurring in metropolitan Cape Town are kept at this mortuary. The remaining cases are dealt with at the Tygerberg hospital mortuary. It is believed, however, that the exclusion of the Tygerberg data does not materially affect the conclusions.

The data consist of all the homicide and suicide cases appearing in the deaths registers relating to the six-year period from 1 January 1986 to 31 December 1991. In each such case the following information was recorded: the deaths register reference, the date of death, the sex of the deceased, a racial classification (African, 'coloured' or white), and the cause of death. The five (mutually exclusive) categories used for the cause of death were: firearm homicide, nonfirearm homicide, firearm suicide, nonfirearm suicide, and 'legal intervention homicide'. (This last category refers to homicide by members of the police or army in the course of their work: in what follows, the word homicide, if unqualified, means homicide other than that resulting from such legal intervention.) Clearly some of the information recorded in the deaths registers could be inaccurate, e.g. a homicide recorded as a suicide, or a legal intervention homicide recorded as belonging to another category. This has to be borne in mind in drawing conclusions from the data.

4.10.1 Models for firearm homicides as a proportion of all homicides, suicides and legal intervention homicides

One question of interest that was examined by means of hidden Markov models was whether there is an upward trend in the pro-

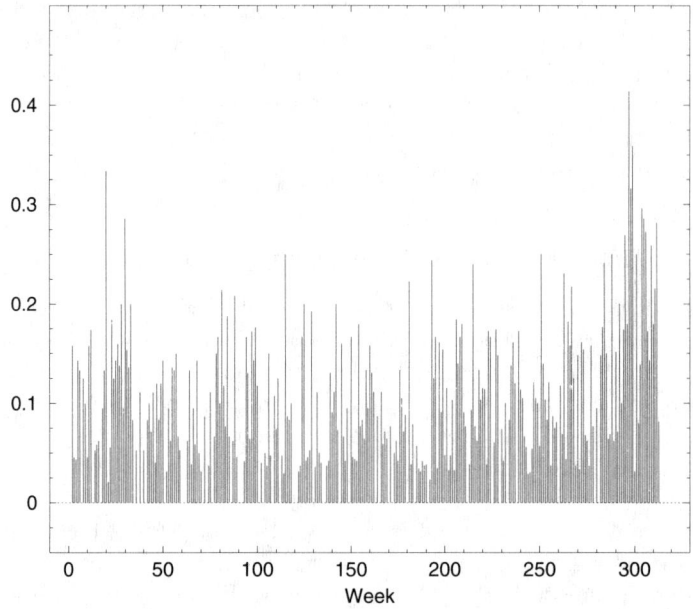

Figure 4.10. *Firearm homicides 1986–1991, as a proportion of all homicides, suicides and legal intervention homicides.*

portion of all the deaths recorded that are firearm homicides. This is of course quite distinct from the question of whether there is an upward trend in the *number* of firearm homicides. The latter kind of trend could be caused by an increase in the population exposed to risk of death, without there being any other relevant change. This distinction is important because of the rapid urbanization which has recently taken place in South Africa and has caused the population in and around Cape Town to increase dramatically.

Four models were fitted to the 313 weekly totals of firearm homicides (given the weekly totals of all the deaths recorded): for a plot of the firearm homicides in each week, as a proportion of all homicides, suicides and legal intervention homicides, see Figure 4.10.

The four models are: a two-state binomial-hidden Markov model with constant 'success probabilities' p_1 and p_2; a similar model with a linear time trend (the same for both states) in the logits of those probabilities; a model allowing differing time-trend parameters in the two states; and finally, a model which assumes that the success

Table 4.31. *Comparison of several binomial-hidden Markov models fitted to the weekly counts of firearm homicides, given the weekly counts of all homicides, suicides and legal intervention homicides.*

model with	k	−l	AIC	BIC
p_1 and p_2 constant	4	590.26	1188.52	1203.50
one time-trend parameter	5	584.34	1178.67	1197.40
two time-trend parameters	6	581.87	1175.75	1198.23
change-point at time 287	6	573.27	**1158.55**	**1181.03**

probabilities are piecewise constant with a single change-point at time 287 — i.e. on 2 July 1991, 26 weeks before the end of the six-year period studied. The time of the change-point was chosen because of the known upsurge of violence in some of the areas adjacent to Cape Town, in the second half of 1991. Much of this violence was associated with the 'taxi wars', a dispute between rival groups of public transport operators.

The models were compared on the basis of AIC and BIC. The results are shown in Table 4.31. Broadly, the conclusion from BIC is that a single (upward) time trend is better than either no trend or two trend parameters, but the model with a change-point is the best of the four. The details of this model are as follows. The underlying Markov chain has transition probability matrix

$$\begin{pmatrix} 0.658 & 0.342 \\ 0.254 & 0.746 \end{pmatrix}$$

and stationary distribution $(0.426, 0.574)$. The probabilities p_1 and p_2 are given by $(0.050, 0.116)$ for weeks 1–287, and by $(0.117, 0.253)$ for weeks 288–313. From this it appears that the proportion of the deaths that are firearm homicides was substantially greater by the second half of 1991 than it was earlier, and that this change is better accommodated by a discrete shift in the probabilities p_1 and p_2 than by gradual movement with time, at least gradual movement of the kind incorporated into the models with time trend. (In passing, this use of a discrete shift further illustrates the flexibility of hidden Markov models.) One further model was also fitted: a model with change-point at the end of week 214. That week included 2 February 1990, on which day President de Klerk made a speech which is widely regarded as a watershed in South Africa's recent history. That model yielded a log-likelihood of −579.83, and AIC and BIC values of 1171.67 and 1194.14. Such a model is therefore superior to the models with time trend, but inferior to the model with the change-point at the end of week 287.

4.10.2 Models for the number of firearm homicides

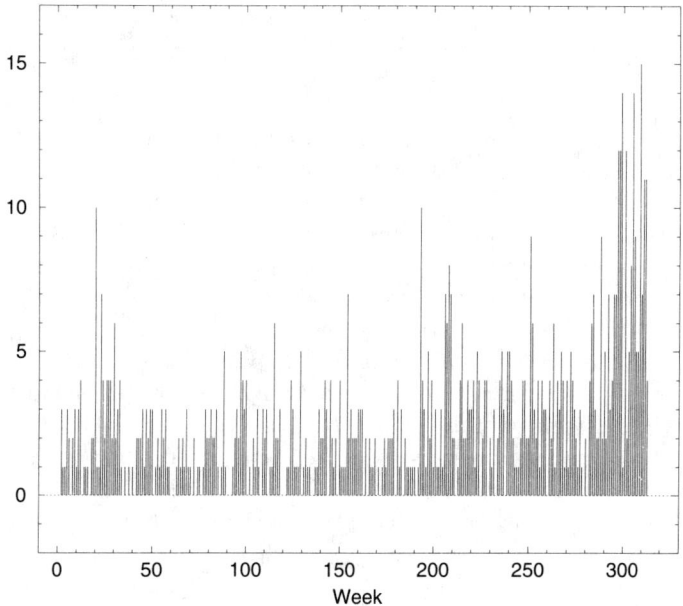

Figure 4.11. *Weekly counts of firearm homicides, 1986–1991.*

In order to model the number (rather than the proportion) of firearm homicides, Poisson-hidden Markov models were also fitted. Weekly counts of firearm homicides are shown in Figure 4.11. The four models fitted in this case were: a two-state model with constant conditional means λ_1 and λ_2; a similar model with a single linear trend in the logs of those means; a model with a quadratic trend therein; and finally, a model allowing for a change-point at time 287. A comparison of these models is shown in Table 4.32.

The conclusion is that, of the four models, the model with a quadratic trend in the conditional means is best. In detail, that model is as follows. The underlying Markov chain has transition

Table 4.32. *Comparison of several Poisson-hidden Markov models fitted to weekly counts of firearm homicides.*

model with	k	−l	AIC	BIC
λ_1 and λ_2 constant	4	626.64	1261.27	1276.26
loglinear trend	5	606.82	1223.65	1242.38
log-quadratic trend	6	602.27	**1216.55**	**1239.02**
change-point at time 287	6	605.56	1223.12	1245.60

probability matrix given by $\begin{pmatrix} 0.881 & 0.119 \\ 0.416 & 0.584 \end{pmatrix}$ and stationary distribution (0.777, 0.223). The conditional means are given by

$$\log \lambda_1 = 0.4770 - 0.004858t + 0.00002665t^2,$$

where t is the week number, and

$$\log \lambda_2 = 1.370 - 0.004858t + 0.00002665t^2.$$

The fact that this smooth trend works better here than does a discrete shift may possibly be explained by population increase due to migration, especially towards the end of the six-year period.

4.10.3 Firearm homicides as a proportion of all homicides, and firearm suicides as a proportion of all suicides

A question of interest that arises from the apparently increased proportion of firearm homicides is whether there is any similar tendency in respect of suicides. Here the most interesting comparison is between firearm homicides as a proportion of all homicides and firearm suicides as a proportion of all suicides. Plots of these two proportions appear as Figures 4.12 and 4.13. Binomial-hidden Markov models of several types were used to model these proportions, and the results are given in Tables 4.33 and 4.34.

The chosen models for these two proportions are as follows. For the firearm homicides the Markov chain has transition probability matrix $\begin{pmatrix} 0.695 & 0.305 \\ 0.283 & 0.717 \end{pmatrix}$ and stationary distribution (0.481, 0.519). The probabilities p_1 and p_2 are given by (0.060, 0.140) for weeks 1–287, and by (0.143, 0.283) for weeks 288–313. The unconditional probability that a homicide involved the use of a firearm is therefore 0.102 before the change-point, and 0.216 there-

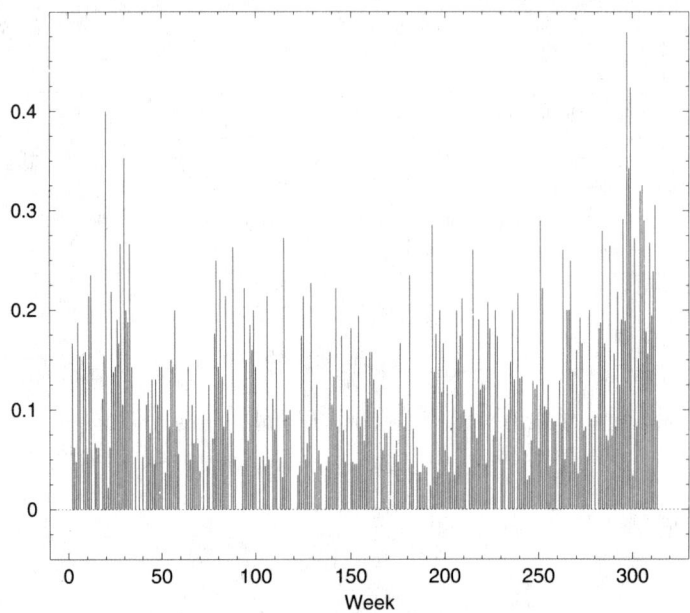

Figure 4.12. *Firearm homicides as a proportion of all homicides, 1986–1991. The week ending on 2 July 1991 is week 287.*

Table 4.33. *Comparison of binomial-hidden Markov models for firearm homicides given all homicides.*

model with	k	$-l$	AIC	BIC
p_1 and p_2 constant	4	590.75	1189.49	1204.48
one time-trend parameter	5	585.59	1181.17	1199.90
two time-trend parameters	6	583.98	1179.95	1202.43
change-point at time 287	6	575.04	**1162.07**	**1184.55**

after. For the firearm suicides, the transition probability matrix is $\begin{pmatrix} 0.854 & 0.146 \\ 0.117 & 0.883 \end{pmatrix}$, and the stationary distribution is (0.446, 0.554). The probabilities p_1 and p_2 are given by (0.186, 0.333), and the unconditional probability that a suicide involves a firearm is 0.267.

A question worth considering, however, is whether time series models are needed at all for these proportions. Is it not perhaps

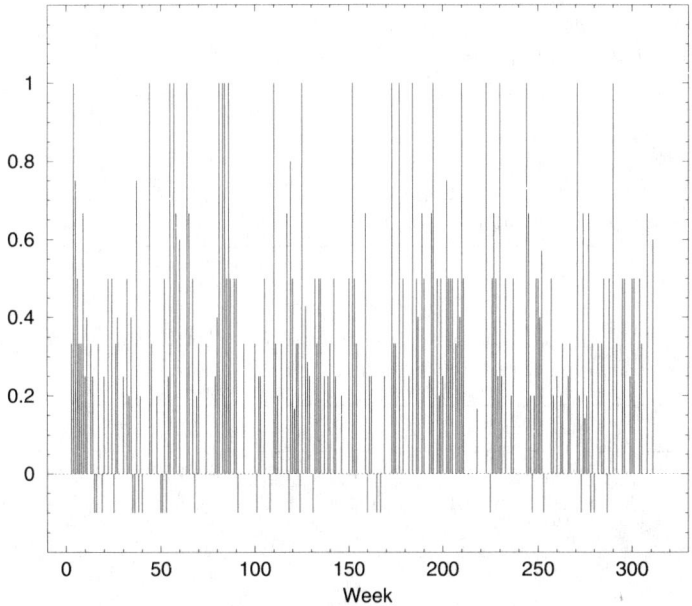

Figure 4.13. *Firearm suicides as a proportion of all suicides, 1986–1991. (Negative values indicate no suicides in that week.)*

Table 4.34. *Comparison of binomial-hidden Markov models for firearm suicides given all suicides.*

model with	k	$-l$	AIC	BIC
p_1 and p_2 constant	4	289.93	**587.86**	**602.84**
one time-trend parameter	5	289.22	588.45	607.18
two time-trend parameters	6	288.30	588.61	611.09
change-point at time 287	6	289.21	590.42	612.90

sufficient to fit a one-state model, i.e. a model which assumes independence of the consecutive observations but is otherwise identical to one of the time series models described above? Models of this type were therefore fitted both to the firearm homicides as a proportion of all homicides, and to the firearm suicides as a proportion of all suicides. The comparison of models is presented in Tables 4.35 and 4.36, in which the parameter p represents the probability that

Table 4.35. *Comparison of several 'independence' models for firearm homicides as a proportion of all homicides.*

model with	k	$-l$	AIC	BIC
p constant	1	637.46	1276.9	1280.7
time trend in p	2	617.80	1239.6	1247.1
change-point at time 287	2	590.60	**1185.2**	**1192.7**

Table 4.36. *Comparison of several 'independence' models for firearm suicides as a proportion of all suicides.*

model with	k	$-l$	AIC	BIC
p constant	1	291.17	**584.33**	**588.08**
time trend in p	2	290.28	584.55	592.04
change-point at time 287	2	291.04	586.09	593.58

a death involves a firearm.

The conclusions that may be drawn from these models are as follows. For the homicides, the models based on independence are without exception clearly inferior to the corresponding hidden Markov time series models. There is sufficient serial dependence present in the proportion of the homicides involving a firearm to render inappropriate any analysis based on an assumption of independence. For the suicides the situation is reversed: the models based on independence are in general superior. There is in this case no convincing evidence of serial dependence, and time series models do not appear to be necessary. The 'best' model based on independence assigns a value of 0.268 to the probability that a suicide involves the use of a firearm, which is of course quite consistent with the value (0.267) implied by the chosen hidden Markov model.

To summarize the conclusions, therefore, we may say that the proportion of homicides that involve firearms does indeed seem to be at a higher level after June 1991, but that there is no evidence of a similar upward shift (or trend) in respect of the proportion of the suicides that involve firearms. There is evidence of serial dependence in the proportion of homicides that involve firearms, but not in the corresponding proportion of suicides. Some further discussion of these findings appears in MacDonald and Lerer (1994).

In view of the finding that the proportion of homicides due to firearms seems to be higher after June 1991, it is interesting to see whether the monitoring technique introduced in section 2.8 would have detected such a change if used over the final two years of the study period. The data for weeks 1–209 only (essentially the first four years, 1986–1989) were used to derive a model with constant probabilities p_1 and p_2 for the weekly numbers of firearm homicides as a proportion of all homicides. For each r from 210 to 313 the conditional distribution (under this model) of S_r given the full history $S^{(r-1)}$ was then computed, and the extremeness or otherwise of the observation s_r was assessed with the help of a plot of forecast pseudo-residuals, which is shown as Figure 4.14.

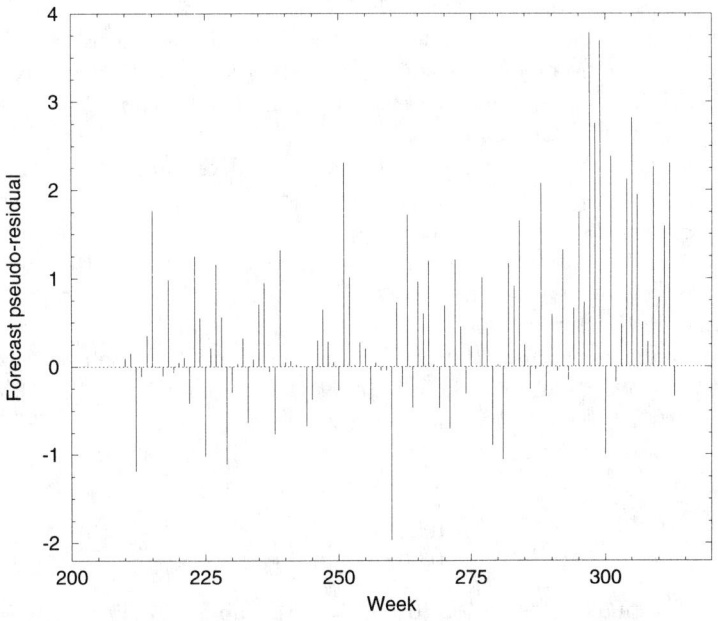

Figure 4.14. *Firearm homicides as a proportion of all homicides: forecast pseudo-residuals from a model based on the data for weeks 1–209 only.*

The result is clear. Not one of the 78 weeks 210–287 produced a tail probability (upper or lower) less than 0.01, but weeks 297, 298, 299, 301 and 305 all produced small or very small conditional probabilities of equalling or exceeding the observed number of firearm homicides: 0.00008, 0.003, 0.0001, 0.009 and 0.002 respectively. (The smallest lower-tail probability in weeks 288–313 was 0.16.) We can therefore conclude that, although the data for 1990 and the first half of 1991 are reasonably consistent with a model based on the four years 1986–1989, the data for the second half of 1991 are not: after June 1991 firearm homicides are at a higher level, relative to all homicides, than is consistent with the model.

4.10.4 Models for the proportions in each of the five categories of death

As a final illustration of the application to these data of hidden Markov models, we describe here two multinomial-hidden Markov models for the weekly totals in each of the five categories of death. These models are of the kind introduced in section 3.3. Each has two states. One model has constant 'success probabilities' and the other allows for a change in these probabilities at time 287. The model without change-point has ten parameters: two to determine the Markov chain, and four independently determined probabilities for each of the two states. The model with change-point has 18 parameters, since there are eight independent probabilities relevant to the period before the change-point, and eight after. For the model without change-point, $-l$ (apart from the constant term involving the multinomial coefficients) is 6463.7, and for the model with change-point it is 6429.3. The corresponding AIC and BIC values are 12 947.5 and 12 985.0 (without change-point), and 12 894.6 and 12 962.0 (with). The model with the change-point at time 287 is therefore preferred, and we give it in full here. The underlying Markov chain has transition probability matrix

$$\begin{pmatrix} 0.541 & 0.459 \\ 0.097 & 0.903 \end{pmatrix}$$

and stationary distribution (0.174, 0.826). Table 4.37 displays, for the period up to the change-point and the period thereafter, the probability of each category of death in state 1 and in state 2, and the corresponding unconditional probabilities. The most noticeable difference between the period before the change-point and the period thereafter is the sharp increase in the unconditional probability of category 1 (firearm homicide), with corresponding decreases in all the other categories.

Clearly the above discussion does not attempt to pursue all the questions of interest arising from these data that may be answered

Table 4.37. *Multinomial-hidden Markov model with change-point at time 287. Probabilities associated with each category of death, before and after the change-point.*

Weeks 1–287

	category 1	2	3	4	5
in state 1	0.124	0.665	0.053	0.098	0.059
in state 2	0.081	0.805	0.024	0.074	0.016
unconditional	0.089	0.780	0.029	0.079	0.023

Weeks 288–313

	category 1	2	3	4	5
in state 1	0.352	0.528	0.010	0.075	0.036
in state 2	0.186	0.733	0.019	0.054	0.008
unconditional	0.215	0.697	0.018	0.058	0.013

Categories:
1. firearm homicide
2. nonfirearm homicide
3. firearm suicide
4. nonfirearm suicide
5. legal intervention homicide.

by the fitting of hidden Markov (or other) time series models. It is felt, however, that the models described, and the conclusions that may be drawn, are sufficiently illustrative of the technique to make clear its utility in such an application.

4.11 Conclusion

Many practical problems involving continuous-valued time series can be approached by a unified strategy based on one class of models, the Gaussian ARMA models. For discrete-valued series, on the other hand, a wider range of models and techniques seems necessary at present.

This chapter has explored the use of one particular class of possible models in a variety of fields of application: medical and sociological applications, finance, animal behaviour, geophysics and climatology. Hidden Markov models of widely differing kinds have been illustrated: models for bounded and for unbounded counts, models with and without time trend, models with change-points, multivariate models, models for categorical series, models with cyclical components and models with a second-order Markov chain as parameter process. From Chapter 3 it will be apparent that the selection of applications presented does not by any means exhaust the variations and techniques that are possible.

The versatility and flexibility of hidden Markov models notwithstanding, it needs to be kept in mind that there will be applications for which observation-driven models are more appropriate than parameter-driven. An observation-driven model may for instance be easier to interpret in the context of a particular application, or it may simply provide a better fit to the data. The mixture transition distribution models of Raftery (section 1.3) and the Markov regression models of Zeger and Qaqish (section 1.7) which are, similarly, versatile classes of models, are available for cases in which an observation-driven model is preferred.

Given the range of discrete-valued models that are available, we believe there is little reason to use normal-theory models for discrete-valued series, except when analysing counts that are large enough to justify a normal approximation, and no reason to regard time series analysis as almost synonymous with the use of Gaussian ARMA models.

There are of course important statistical questions relating to the use of hidden Markov models which need further study: *inter alia* model selection, the properties of maximum likelihood estimators, and the question of testing for goodness of fit. Some of these may be difficult problems: Reeves (1993) observes, in the simpler context of testing the goodness of fit of Markov chain models for binary time series, that most methods for analysing binary data assume independence and do not easily extend to the case of dependence.

It is, however, hoped that the account of the models and their properties presented in Chapters 2 and 3, and the illustrative applications presented in this chapter, will persuade the reader that hidden Markov models are a useful addition to the techniques available to the statistician who meets discrete-valued time series in her or his work.

APPENDIX A

Proofs of results used in the derivation of the Baum–Welch algorithm

The purpose of this appendix is to derive properties (2.1)–(2.4), which were stated without proof in section 2.2 and used there in the derivation of the Baum–Welch algorithm. All four results refer to the following situation. The processes $\{C_t : t \in \mathbf{N}\}$ and $\{S_t : t \in \mathbf{N}\}$ are finite state-space processes. $\{C_t\}$ is a homogeneous Markov chain and $\{S_t\}$ has (for all T) the property that, conditional on $C^{(T)}$, the random variables $S_1, S_2, \ldots, S_T$ are mutually independent and the (conditional) distribution of S_t is given by $\mathrm{P}(S_t \mid C_t)$. That is, this distribution depends only on C_t and not on any C_k, $k \neq t$.

The technique of proof is in general as follows:

(a) express the probability of interest in terms of probabilities conditional on $C^{(T)}$, i.e. conditional on all of $C_1, \ldots, C_T$;

(b) use the fact that, conditional on $C^{(T)}$, the random variables $S_1, \ldots, S_T$ are independent, with the distribution of each S_t depending only on the corresponding C_t;

(c) use the Markov property of $\{C_t\}$ if necessary.

We establish first property (2.1), and then derive property (2.4) from it. To establish property (2.1) we use Propositions 1, 2 and 3 given below. Properties (2.2) and (2.3) then follow.

Proposition 1 *For all integers t and l such that $1 \leq t \leq l \leq T$:*

$$\mathrm{P}(S_l, S_{l+1}, \ldots, S_T \mid C_t, \ldots, C_T) = \mathrm{P}(S_l, \ldots, S_T \mid C_l, \ldots, C_T).$$

Proof. The left-hand side above can be written as:

$$\frac{1}{\mathrm{P}(C_t, \ldots, C_T)} \sum_{c_1, \ldots, c_{t-1}} \mathrm{P}(S_l, \ldots, S_T \mid C^{(T)})\, \mathrm{P}(C^{(T)}),$$

there being no summation in the case $t=1$. By (b) we have
$$P(S_l,\ldots,S_T \mid C^{(T)}) = P(S_l \mid C_l) \cdots P(S_T \mid C_T),$$
which can be taken outside the summation. The resulting sum then reduces to $P(C_t,\ldots,C_T)$, and the left-hand side is seen to be just
$$P(S_l \mid C_l) \cdots P(S_T \mid C_T),$$
which is independent of t. The right-hand side, being the case $t=l$ of the left-hand side, equals the same expression. $\square$

Proposition 2 *For $t = 1, 2, \ldots, T-1$:*
$$P(S_{t+1},\ldots,S_T \mid C^{(t)}) = P(S_{t+1},\ldots,S_T \mid C_t).$$

Proof. The left-hand side can be written as
$$\frac{1}{P(C^{(t)})} \sum_{c_{t+1},\ldots,c_T} P(C^{(T)}) \, P(S_{t+1},\ldots,S_T \mid C^{(T)}).$$

Now apply Proposition 1 (twice) and the Markov property of $\{C_t\}$ to see that this equals
$$\sum_{c_{t+1},\ldots,c_T} P(C_{t+1},\ldots,C_T \mid C_t) \, P(S_{t+1},\ldots,S_T \mid C_t,\ldots,C_T).$$

The summand is $P(S_{t+1},\ldots,S_T,C_t,\ldots,C_T)/P(C_t)$, and the sum is therefore equal to $P(S_{t+1},\ldots,S_T,C_t)/P(C_t)$, as required. $\square$

Proposition 3 *For $t = 1, 2, \ldots, T$:*
$$P(S_1,\ldots,S_t \mid C^{(T)}) = P(S_1,\ldots,S_t \mid C^{(t)}).$$

Proof. Apply (b) in respect of the conditioning on $C^{(T)}$ to see that the left-hand side equals $P(S_1 \mid C_1) \cdots P(S_t \mid C_t)$. Apply (b) in respect of the conditioning on $C^{(t)}$ to see that the right-hand side equals the same expression. $\square$

Proposition 4 (Property (2.1)) *For $t = 1, 2, \ldots, T$:*
$$P(S_1,\ldots,S_T \mid C_t) = P(S_1,\ldots,S_t \mid C_t) \, P(S_{t+1},\ldots,S_T \mid C_t).$$

Proof. Making use of the mutual independence of $S_1,\ldots,S_T$ given $C^{(T)}$, write the left-hand side as
$$\frac{1}{P(C_t)} {\sum}^{(1)} {\sum}^{(2)} P(C^{(T)})$$
$$\times \left(P(S_1,\ldots,S_t \mid C^{(T)}) \, P(S_{t+1},\ldots,S_T \mid C^{(T)}) \right),$$

RESULTS USED IN THE BAUM–WELCH ALGORITHM

where $\sum^{(1)}$ denotes summation over $c_1, \ldots, c_{t-1}$, and $\sum^{(2)}$ over $c_{t+1}, \ldots, c_T$. Then apply Propositions 3 and 2 to show that this equals

$$\frac{1}{\mathrm{P}(C_t)} \left(\sum\nolimits^{(1)} \mathrm{P}(S_1, \ldots, S_t, C_1, \ldots, C_t) \right) \mathrm{P}(S_{t+1}, \ldots, S_T \mid C_t)$$
$$= \frac{1}{\mathrm{P}(C_t)} \mathrm{P}(S_1, \ldots, S_t, C_t) \, \mathrm{P}(S_{t+1}, \ldots, S_T \mid C_t),$$

i.e. the right-hand side. □

Proposition 5 (Property (2.4)) *For $t = 1, 2, \ldots, T$:*

$$\mathrm{P}(S_t, \ldots, S_T \mid C_t) = \mathrm{P}(S_t \mid C_t) \, \mathrm{P}(S_{t+1}, \ldots, S_T \mid C_t).$$

Proof. Sum the result of Proposition 4 with respect to $s_1, \ldots, s_{t-1}$. □

Proposition 6 (Property (2.2)) *For $t = 1, 2, \ldots, T-1$:*

$$\mathrm{P}(S_1, \ldots, S_T \mid C_t, C_{t+1})$$
$$= \mathrm{P}(S_1, \ldots, S_t \mid C_t) \, \mathrm{P}(S_{t+1}, \ldots, S_T \mid C_{t+1}).$$

Proof. Write the left-hand side as

$$\frac{1}{\mathrm{P}(C_t, C_{t+1})} \sum\nolimits^{(1)} \sum\nolimits^{(2)} \mathrm{P}(C^{(T)})$$
$$\times \left(\mathrm{P}(S_1, \ldots, S_t \mid C^{(T)}) \, \mathrm{P}(S_{t+1}, \ldots, S_T \mid C^{(T)}) \right),$$

where $\sum^{(1)}$ denotes summation over $c_1, \ldots, c_{t-1}$ and $\sum^{(2)}$ over $c_{t+2}, \ldots, c_T$. By Propositions 3 and 1 respectively, the last two factors in the above expression reduce to $\mathrm{P}(S_1, \ldots, S_t \mid C^{(t)})$ and $\mathrm{P}(S_{t+1}, \ldots, S_T \mid C_{t+1}, \ldots, C_T)$. The Markov property of $\{C_t\}$ is then used, and after some routine manipulations of conditional probabilities it emerges that the above expression is equal to

$$\mathrm{P}(S_1, \ldots, S_t \mid C_t) \frac{1}{\mathrm{P}(C_{t+1})} \sum\nolimits^{(2)} \mathrm{P}(S_{t+1}, \ldots, S_T, C_{t+1}, \ldots, C_T)$$
$$= \mathrm{P}(S_1, \ldots, S_t \mid C_t) \, \mathrm{P}(S_{t+1}, \ldots, S_T, C_{t+1}) / \mathrm{P}(C_{t+1}),$$

as required. □

Proposition 7 (Property (2.3)) *For all integers t and l such that $1 \leq t \leq l \leq T$:*

$$\mathrm{P}(S_l, \ldots, S_T \mid C_t, \ldots, C_l) = \mathrm{P}(S_l, \ldots, S_T \mid C_l).$$

Proof. If $\sum^{(1)}$ denotes summation over $c_1, \ldots, c_{t-1}$ and $\sum^{(2)}$ over $c_{l+1}, \ldots, c_t$, the left-hand side can be written as

$$\frac{1}{P(C_t, \ldots, C_l)} \sum\nolimits^{(2)} \sum\nolimits^{(1)} P(S_l, \ldots, S_T \mid C^{(T)}) \, P(C^{(T)}).$$

By Proposition 1

$$P(S_l, \ldots, S_T \mid C^{(T)}) = P(S_l, \ldots, S_T \mid C_1, \ldots, C_T),$$

and the above expression for the left-hand side therefore equals

$$\sum\nolimits^{(2)} P(S_l, \ldots, S_T \mid C_1, \ldots, C_T) \, P(C_{l+1}, \ldots, C_T \mid C_t, \ldots, C_l).$$

By the Markov property of $\{C_t\}$, this equals

$$\sum\nolimits^{(2)} P(S_l, \ldots, S_T \mid C_1, \ldots, C_T) \, P(C_{l+1}, \ldots, C_T \mid C_l)$$
$$= \frac{1}{P(C_l)} \sum\nolimits^{(2)} P(S_l, \ldots, S_T, C_1, \ldots, C_T)$$
$$= P(S_l, \ldots, S_T, C_l)/P(C_l),$$

as required. □

APPENDIX B

Data

In this appendix we give seven of the nine data sets discussed in Chapter 4. The exceptions are the evapotranspiration data, which are given in the text in section 4.7, and the lengthy series of wind directions discussed in section 4.6. It is intended that all the data analysed in this book, as well as the Fortran programs used to do so, will be available by anonymous ftp from directory /data/pub/h-markov at the following server at the University of Göttingen: wso140serv.wiso.gwdg.de. These programs do, however, require that several of the routines in the NAG Fortran Library (Numerical Algorithms Group, 1992) are available.

Old Faithful geyser

Table B.1, which is to be read across rows, contains data on the duration of 299 successive eruptions of the Old Faithful geyser during the period 1–15 August 1985. Eruptions of short duration are denoted by 0, and long by 1. For a fuller description of the data, see section 4.2. The data are derived from Table 1 of Azzalini and Bowman (1990). The Royal Statistical Society is thanked for permission to print these data.

Table B.1. *Short and long eruptions of Old Faithful geyser*

```
1 0 1 1 1 0 1 1 0 1 0 1 0 1 1 0 1 0 1 1 0 1 0 1 0 1 0 1 1 1
1 1 0 1 0 1 0 1 0 1 0 1 0 1 0 1 0 1 0 1 0 1 0 1 1 1 1 1
0 1 0 1 0 1 0 1 1 0 1 0 1 1 1 0 1 1 1 1 0 1 1 1 0 1 0 1 0
1 0 1 0 1 0 1 0 1 0 1 0 1 0 1 0 1 0 1 1 0 1 0 1 0 1 0 1
0 1 1 1 0 1 1 1 1 1 1 0 1 1 1 1 0 1 1 1 1 1 1 0 1 0 1
0 1 0 1 0 1 1 1 1 1 0 1 0 1 0 1 0 1 1 1 0 1 0 1 0 1 1
0 1 0 1 1 1 1 0 1 0 1 0 1 1 1 0 1 0 1 0 1 1 0 1 1 0 1 1
1 0 1 0 1 0 1 0 1 1 0 1 1 1 1 1 1 0 1 0 1 0 1 1 1 1 0 1 1
0 1 1 1 0 1 1 0 1 0 1 1 1 0 1 0 1 1 1 1 0 1 1 1 0 1 0 1 0
1 1 0 1 0 1 1 1 1 1 1 1 1 0 1 0 1 0 1 0 1 0 1 0 1 0 1 1 0
```

Epileptic seizures

Table B.2, also to be read across rows, consists of the numbers of myoclonic seizures suffered by one patient on each of 225 consecutive days. The data were read from Figure 1 of Le, Leroux and Puterman (1992). For further background information, see that paper.

We have, however, been informed that observations 92–112 (i.e. weeks 14–16) were apparently included in error and should be deleted. These 21 observations appear in italics in the table. The corrected series is analysed in section 4.3.

We are grateful to Dr Nhu Le for his assistance in identifying the observations that need to be corrected.

Table B.2. *Epileptic seizure counts*

```
0 3 0 0 0 0 1    1 0 2 1 1 2 0    0 1 2 1 3 1 3
0 4 2 0 1 1 2    1 2 1 1 1 0 1    0 2 2 1 2 1 0
0 0 2 1 2 0 1    0 1 0 1 0 0 0    0 0 0 0 0 1 0
0 0 0 0 1 0 0    0 1 0 0 0 1 0    0 0 1 0 0 1 0
0 2 1 0 1 1 0    0 0 2 2 0 1 1    3 1 1 2 1 0 3
6 1 3 1 2 2 1    0 0 2 2 0 1 1    3 1 1 2 1 0 3
6 1 3 1 2 2 1    0 1 2 1 0 1 2    0 0 2 2 1 0 1
0 0 2 0 1 0 0    0 1 0 0 1 0 0    0 0 0 0 0 1 3
0 0 0 0 0 1 0    1 1 1 0 0 0 0    0 1 0 1 2 1 0
0 0 0 0 0 1 4    0 0 0 0 0 0 0    0 0 0 0 0 0 0
0 0 0 0 0 0 0    0 0 0 0 0 0 0    0
```

Edendale births

Table B.3 presents the total number of births, and the number of deliveries by various methods, in the Obstetrics Unit of Edendale Hospital, Natal, South Africa, during each month of the nine-year period from February 1977 to January 1986. The table is to be read as follows: first read down the first block of five columns, then similarly down the second and third blocks of five columns. For each month there are the following five items, in order: the number of mothers delivered, the number of Caesarean sections performed, the number of breech births, the number of forceps deliveries, and the number of vacuum extraction deliveries.

Prof. Linda Haines is thanked for making these data available.

Table B.3. *Edendale births*

553	87	6	35	35	523	130	27	38	13	711	185	14	18	43
612	96	14	29	27	540	127	15	36	14	760	234	17	23	55
635	104	9	30	41	533	156	21	32	7	764	225	11	13	55
671	110	18	58	51	534	149	26	26	17	741	249	19	15	45
692	107	39	82	50	553	144	35	29	20	673	196	19	19	53
672	124	24	86	49	636	150	11	35	22	743	178	8	35	60
690	122	31	70	31	604	158	15	38	26	775	222	7	23	44
767	161	17	56	46	670	168	25	32	34	896	221	18	28	85
627	130	5	48	35	627	152	27	26	34	727	192	20	27	45
586	117	0	46	44	532	101	25	15	16	649	173	17	9	28
691	138	17	32	48	565	132	26	19	18	722	185	14	12	50
645	131	12	32	38	562	124	14	52	27	680	196	10	13	51
594	141	34	24	29	544	127	9	46	36	716	208	7	19	68
594	117	10	40	43	602	154	7	44	32	755	225	15	14	48
607	134	30	42	20	582	126	35	39	58	736	212	10	9	46
658	122	10	61	14	615	127	16	42	51	740	198	19	9	40
559	118	12	34	19	606	152	15	34	47	780	204	24	3	43
614	125	21	44	21	643	157	25	23	32	743	193	16	15	38
673	146	18	56	20	672	145	15	32	43	864	219	22	23	54
753	168	21	62	23	768	191	20	37	41	921	249	17	18	60
656	161	25	47	16	679	178	15	48	37	820	234	6	19	28
657	141	24	37	11	629	147	25	37	35	777	207	15	30	18
648	167	31	44	20	719	188	15	45	36	847	210	23	42	17
558	124	26	37	23	685	177	6	57	36	817	207	18	32	45
571	152	16	58	15	591	154	10	37	21	773	192	27	17	60
626	155	13	77	4	651	145	10	18	26	883	209	26	27	66
571	102	27	81	28	622	172	14	23	32	777	205	13	16	53
618	167	30	67	32	703	171	10	28	34	826	201	14	24	51
584	147	23	58	16	661	168	12	46	23	784	223	19	19	54
605	133	24	58	14	714	210	10	26	56	824	222	25	20	61
659	142	11	64	10	776	227	14	28	58	853	258	19	8	64
687	135	36	81	15	869	239	6	28	78	970	287	14	26	46
615	157	32	55	3	742	202	8	36	51	831	261	17	20	35
592	126	25	51	10	699	198	24	19	46	744	221	18	14	41
552	120	22	41	11	710	183	21	16	72	817	207	15	14	49
598	130	31	35	18	710	197	19	19	53	780	186	18	12	37

Locomotion of locusts

Table B.4 contains, in hexadecimal form, 161 consecutive binary observations on each of 24 locusts, in an experiment performed by Dr David Raubenheimer. For example, the first eight observations on subject 10 are 1101 0010, the binary equivalent of D2. The observation is 1 if the locust was locomoting at the relevant time, otherwise 0. (For the purpose of the conversion from binary to hexadecimal, the 24 series have all been packed with zeros in positions 162–164.) For a fuller description of the experiment, see section 4.5.

Table B.4. *Locomotion of locusts*

1– 4	0 0 0 4	0 0 0 5	0 D 0 0	0 0 0 0	0 0 0 4	0 0 0 0
5– 8	0 0 2 8	0 8 0 8	0 2 0 0	0 0 0 0	0 0 0 4	0 2 0 0
9– 12	0 0 1 E	0 3 0 F	0 B 0 0	0 0 0 0	4 0 1 0	2 0 1
13– 16	0 0 0 0	0 2 6 B	0 F 0 0	0 8 0 0	0 4 0 7	0 0 0 0
17– 20	0 0 0 0	0 F 0 4	0 4 0 0	0 1 0 0	0 0 0 9	0 4 0 0
21– 24	0 3 0 4	8 2 0 0	0 F 0 0	0 6 0 0	0 5 0 5	0 1 0 0
25– 28	0 A 0 6	0 7 0 5	4 C 0 0	0 0 2 0	0 0 0 7	1 9 0 0
29– 32	0 F 0 0	0 F 0 2	0 5 0 1	0 F 0 0	0 1 0 4	4 E 0 0
33– 36	0 C 0 6	0 F 0 7	0 F 4 1	0 8 0 0	0 0 0 4	3 0 0 0
37– 40	0 F 0 B	0 F 0 D	0 9 0 0	0 1 0 0	0 E 0 B	4 4 1 0
41– 44	0 F 1 7	0 F 0 F	0 A 0 0	3 4 0 8	0 4 0 4	8 0 0 C
45– 48	0 4 0 7	0 B 0 F	0 A 0 0	0 0 0 C	0 1 0 E	0 1 0 4
49– 52	0 2 0 D	0 F 0 F	0 A 0 0	0 1 0 0	0 1 0 6	0 B 0 E
53– 56	0 D 0 B	0 D 0 E	0 3 0 1	0 0 0 5	0 4 0 F	0 9 0 1
57– 60	0 E 0 4	0 F 0 F	0 C 0 B	0 0 0 B	0 7 0 C	0 8 0 0
61– 64	0 D 4 3	0 E 0 F	0 1 0 3	0 1 0 2	0 7 0 E	0 0 0 3
65– 68	0 1 0 F	0 F 0 B	0 D 0 6	C B 0 C	0 7 0 F	6 2 0 F
69– 72	0 F 0 5	0 F 0 D	0 9 0 5	0 E 2 5	0 E 0 5	2 6 0 8
73– 76	0 C 0 6	1 E 3 7	0 D 0 4	0 B 0 6	0 E 0 8	0 0 0 0
77– 80	0 3 0 3	8 F 4 B	0 E 0 8	0 5 0 C	0 E 0 F	0 0 0 0
81– 84	0 B 0 C	0 F 0 F	0 B 0 B	8 3 0 1	0 4 0 F	4 C 0 0
85– 88	0 C 0 0	0 F 0 F	0 C 0 1	0 F 0 D	0 9 0 7	2 0 0 0
89– 92	0 F 0 2	0 D 0 B	0 B 0 E	0 F 0 9	0 8 0 F	2 3 0 0
93– 96	0 1 0 4	0 F 0 F	0 8 0 0	0 F 0 8	0 0 E F	3 9 0 1
97–100	0 F 0 2	0 F 0 D	0 4 0 2	1 D 0 0	0 5 0 F	0 4 0 C
101–104	0 6 0 9	0 F 0 F	0 C 0 0	1 F 0 0	0 8 0 F	0 7 0 B
105–108	0 3 0 1	0 F 0 F	0 1 0 0	0 7 2 8	0 4 4 D	0 D 0 9
109–112	0 B 1 6	0 E 0 7	0 3 0 1	E F 0 F	0 F 0 F	0 B 0 C
113–116	0 D 0 C	0 F 0 E	0 D 0 0	8 7 0 E	0 0 1 F	0 F 0 6
117–120	8 7 2 A	0 F 0 F	0 B 0 C	5 F 0 5	0 0 4 9	0 0 0 1
121–124	0 4 0 E	1 4 0 F	0 E 0 0	1 E 0 7	0 0 B F	C 1 0 7
125–128	0 0 2 8	8 E 0 E	0 E 0 0	1 7 0 C	0 3 3 F	3 1 0 F
129–132	0 B F 1	1 E 0 F	0 9 0 0	5 F 0 C	0 F C D	0 F 0 D
133–136	0 B 4 B	0 F 0 F	0 7 0 0	D E 0 3	0 C 1 B	4 8 0 9
137–140	0 F F E	0 F 0 F	0 F 0 0	0 F 0 F	2 5 C F	4 E 0 F
141–144	0 F F 1	0 F 0 F	0 B 8 0	1 F 0 1	8 2 C 7	0 6 0 A
145–148	0 E 3 0	0 C 0 D	0 1 0 0	0 F 0 C	0 5 7 7	0 8 0 9
149–152	0 F D C	0 F A 7	B 7 0 D	2 B 0 6	0 8 7 9	4 0 0 F
153–156	0 9 E 0	8 2 1 D	8 E 0 0	0 7 0 C	1 0 F 9	0 4 0 E
157–160	0 4 6 5	0 4 0 F	0 A 0 0	4 7 0 F	7 5 5 A	F 8 0 7
161–161	0 8 8 8	0 0 0 0	0 8 0 0	0 8 0 8	8 0 0 8	8 0 0 8

Thinly traded shares

Table B.5 contains, in hexadecimal form, data for 910 days indicating presence or absence of trading in the six thinly traded shares discussed in section 4.8. The shares are, in order: Amcoal, Vierfontein, Wankie, Anamint, Broadacres and Carrigs. For example, the data for Anamint for the first 4 days are 0111, the binary equivalent of 7. The series have been packed with zeros in positions 911 and 912.

Dr David Bowie is thanked for making these data available.

Table B.5. *Thinly traded shares*

1– 4	F F C 7 F F	125–128	F 7 8 F 8 8	249–252	9 F 0 5 1 1
5– 8	F F 4 E F F	129–132	F 6 0 B 8 0	253–256	6 4 0 C 0 E
9– 12	F 5 5 6 E F	133–136	9 D 4 F 4 3	257–260	F 9 0 6 0 8
13– 16	8 7 0 0 0 7	137–140	F 5 0 D 4 C	261–264	F 8 0 D 0 C
17– 20	C C 0 0 F F	141–144	7 0 0 F 0 7	265–268	D 1 0 1 0 F
21– 24	7 F 3 0 B F	145–148	F A 0 F 1 E	269–272	7 2 F 2 7 D
25– 28	F E 6 4 F F	149–152	7 D 0 E 1 4	273–276	F 0 0 F 8 E
29– 32	F D 1 6 2 F	153–156	7 3 0 D 9 F	277–280	E F 0 6 1 C
33– 36	D E 0 0 C 7	157–160	7 E 0 2 A E	281–284	F E 0 7 7 7
37– 40	A 6 0 0 E F	161–164	F 6 8 F 1 7	285–288	F C 0 C 3 A
41– 44	F D 0 0 6 E	165–168	3 5 1 F 5 D	289–292	F C E C 5 2
45– 48	F 4 8 0 8 F	169–172	5 9 0 4 4 A	293–296	D E 5 1 F C
49– 52	E B 8 6 5 E	173–176	F A 0 D 5 E	297–300	F 0 0 F 2 6
53– 56	E 8 0 2 D F	177–180	D 4 4 5 8 D	301–304	7 F 8 A 2 C
57– 60	D 1 2 1 4 7	181–184	B 3 0 2 3 A	305–308	3 F 1 0 2 1
61– 64	E 1 0 7 C 5	185–188	B D 0 0 6 D	309–312	B F 0 8 7 2
65– 68	F 3 0 C F 7	189–192	F D C 6 8 F	313–316	F F 0 B 9 1
69– 72	F 3 B 3 F 7	193–196	7 F 0 3 A E	317–320	F 7 9 7 F B
73– 76	B 3 4 D 6 E	197–200	E F 0 6 C 3	321–324	E 6 4 3 7 F
77– 80	F E 0 7 7 7	201–204	2 F 1 2 A 6	325–328	7 F 6 3 F F
81– 84	B F 1 0 7 D	205–208	3 3 0 B 3 E	329–332	F F 0 0 B 1
85– 88	F 7 5 0 7 D	209–212	B 9 0 7 7 B	333–336	B F 0 F 6 6
89– 92	D F 1 D 3 A	213–216	C 7 0 F 1 3	337–340	7 D 0 F F F
93– 96	7 D C F 0 3	217–220	F D 0 C 1 7	341–344	E 4 4 0 0 0
97–100	F D 4 7 4 F	221–224	F E 0 F D 6	345–348	D F 1 A A 0
101–104	B D 2 D E E	225–228	D 7 0 E 7 D	349–352	D F 2 D 4 F
105–108	F B 1 7 4 E	229–232	F 7 0 3 F 5	353–356	D F 4 7 9 7
109–112	D D 1 E 0 7	233–236	F 0 0 4 F F	357–360	F F B D D F
113–116	D 1 A F 6 A	237–240	B 3 0 5 A 4	361–364	F B 4 3 C F
117–120	6 0 0 F 3 8	241–244	D 7 2 A A 9	365–368	F 3 0 3 9 F
121–124	E E 8 B C B	245–248	C 3 0 4 0 C	369–372	D 0 0 E F 3

(continued on next page)

Table B.5. *Thinly traded shares (continued)*

Range	Values	Range	Values	Range	Values
373–376	F F 0 4 6 F	553–556	6 1 0 7 1 D	733–736	E 0 0 F 0 F
377–380	A D 0 E C F	557–560	3 9 0 8 2 B	737–740	4 8 0 D 1 C
381–384	E E 1 4 2 7	561–564	A A 0 B 0 3	741–744	B 0 0 7 5 0
385–388	9 7 8 0 0 9	565–568	1 A 9 8 7 F	745–748	4 C 5 F 2 A
389–392	F F 0 4 4 E	569–572	2 3 B 6 5 D	749–752	B 4 4 F 6 0
393–396	D E 0 7 0 D	573–576	C 9 1 6 E F	753–756	D 0 2 F 1 8
397–400	F F 2 9 0 F	577–580	C C 0 6 5 9	757–760	D 8 2 F 0 8
401–404	A 8 0 F 0 F	581–584	F 9 0 B A F	761–764	0 5 4 6 0 0
405–408	F 0 0 C 0 0	585–588	7 F 0 7 6 F	765–768	5 A 8 C 0 0
409–412	C 6 0 4 1 0	589–592	E F 0 6 4 1	769–772	F 0 8 F 0 2
413–416	0 D 0 A 7 0	593–596	F 3 1 6 6 F	773–776	A 0 4 F 0 8
417–420	8 B 0 0 A 0	597–600	5 7 0 5 2 9	777–780	F 4 0 3 C 0
421–424	5 6 0 A 6 0	601–604	E 2 8 7 F 6	781–784	9 0 0 B 0 5
425–428	E 2 0 B 5 0	605–608	E 6 0 F 5 E	785–788	2 8 4 F 0 1
429–432	E 1 D F A 0	609–612	F 8 0 E 2 1	789–792	0 2 4 B 0 8
433–436	D 6 0 3 E 0	613–616	F 9 4 8 C 1	793–796	7 0 0 D 0 8
437–440	D 8 4 E 1 0	617–620	E A 0 7 3 9	797–800	E 5 0 E 0 2
441–444	D 3 0 1 4 0	621–624	D 0 0 C B E	801–804	0 8 2 0 0 1
445–448	7 F 0 F 3 1	625–628	E 9 1 4 F B	805–808	5 0 0 F 0 0
449–452	F B 0 C C F	629–632	2 3 C 9 2 B	809–812	9 0 0 F 0 4
453–456	7 F C 8 8 F	633–636	F C 0 9 6 E	813–816	F 0 2 F 0 2
457–460	B F 0 1 1 F	637–640	B 7 4 9 0 6	817–820	5 4 0 F 0 0
461–464	7 7 5 8 1 F	641–644	9 9 0 F 2 F	821–824	E 5 0 2 6 0
465–468	4 F 0 F B F	645–648	F 3 9 F 0 E	825–828	C 2 0 E 2 1
469–472	F F 6 E B F	649–652	3 0 2 F 1 2	829–832	E 0 0 D 0 E
473–476	7 F 0 2 0 F	653–656	B C 4 6 8 7	833–836	9 0 4 5 5 1
477–480	7 F 0 9 E F	657–660	E 6 0 2 3 2	837–840	8 2 A 7 D 1
481–484	1 1 0 5 8 F	661–664	F E 0 7 4 0	841–844	7 0 0 F 5 2
485–488	0 4 0 8 0 E	665–668	1 D 0 F 1 0	845–848	4 4 0 F 2 0
489–492	E B A 1 A F	669–672	F 2 0 8 4 7	849–852	6 5 0 F 9 0
493–496	7 B 4 A F F	673–676	C 1 0 E 8 B	853–856	6 1 0 D 0 0
497–500	1 2 3 8 0 B	677–680	F 3 9 C 4 D	857–860	A 0 1 7 4 0
501–504	9 8 0 0 9 5	681–684	E 5 1 6 0 7	861–864	4 3 0 E 0 2
505–508	F F 1 0 8 F	685–688	D E 0 9 3 0	865–868	E 0 6 3 B 0
509–512	F D 2 E 3 C	689–692	2 0 2 9 2 2	869–872	C 2 2 B 0 0
513–516	F 6 0 F 0 0	693–696	6 1 1 F 4 0	873–876	7 2 0 F 0 3
517–520	A 3 2 F 4 7	697–700	F 6 0 F 1 9	877–880	6 2 0 F 0 2
521–524	5 F 0 8 1 0	701–704	E E 2 4 8 B	881–884	B 8 0 F 1 7
525–528	D F 0 3 0 9	705–708	7 6 0 7 0 D	885–888	D 3 0 F 1 9
529–532	F F 0 C 8 0	709–712	F 5 0 E 0 E	889–892	F 0 2 F 1 8
533–536	A 9 2 0 8 F	713–716	F 5 2 F 5 F	893–896	E 4 0 F C 9
537–540	D 0 4 C 0 F	717–720	B 5 8 B 0 4	897–900	F 8 0 B 1 0
541–544	F E 3 5 8 D	721–724	B 9 0 F 2 0	901–904	6 F 0 F 0 0
545–548	3 D 0 1 C 9	725–728	5 1 2 3 0 0	905–908	F 1 8 C 8 8
549–552	C C 0 8 8 A	729–732	E 7 8 F 0 D	909–910	4 8 0 C 0 0

Durban daily rainfall

Table B.6 contains the Durban rainfall data, in the form needed to fit the nonhomogeneous Markov chain model described in section 4.9. The tth set of four numbers ($t = 1, 2, 3, \ldots, 365$) consists of the following, in order: $N_{WD}(t)$, $N_{WO}(t)$, $N_{DW}(t)$, and $N_{DO}(t)$. Here $N_{WD}(t)$ is the number of times day $t - 1$ was wet and day t dry; $N_{WO}(t)$ the number of times day $t - 1$ was wet and there was an observation on day t; $N_{DW}(t)$ the number of times day $t - 1$ was dry and day t wet; and $N_{DO}(t)$ the number of times day $t - 1$ was dry and there was an observation on day t.

The layout of the table is as follows: first read down the first block of four columns, then down each of the second to sixth blocks, then read the second page of the table similarly. For instance $N_{WD}(2) = 23$ and $N_{DO}(27) = 63$.

Table B.6. *Daily rainfall at Durban, 1871–1992*

20 39 26 69	18 50 20 63	16 33 29 77	18 36 16 73	25 36 17 75	17 19 16 92
23 46 24 67	26 52 16 61	26 46 17 64	17 34 22 75	20 28 19 83	11 18 16 93
20 47 18 66	18 42 32 71	21 37 26 73	20 39 17 70	10 27 18 84	12 23 13 88
21 45 19 68	32 56 19 57	18 42 26 68	17 36 17 73	22 35 17 76	20 24 18 87
26 43 24 70	13 43 23 70	19 50 14 60	17 36 20 73	20 30 11 81	15 22 10 89
18 41 31 72	29 50 16 60	28 45 20 65	22 39 18 70	13 21 22 90	10 17 11 94
23 54 19 59	19 37 18 73	13 37 23 73	20 35 20 74	18 30 18 81	10 18 15 93
25 50 22 63	14 36 24 74	24 46 21 63	19 35 27 74	19 30 19 81	12 23 13 88
25 47 22 66	27 46 24 64	21 43 28 66	24 43 14 66	16 30 13 81	14 24 10 87
22 44 22 69	20 43 20 67	23 50 20 59	11 33 28 76	20 27 14 84	9 20 12 91
19 44 23 69	21 43 22 67	18 47 20 62	27 50 13 59	12 21 15 90	14 23 16 88
21 48 20 65	18 44 23 66	23 49 15 60	22 36 18 73	16 24 17 87	17 25 13 86
19 47 19 66	21 49 28 61	17 41 17 68	18 32 21 77	17 25 11 86	18 21 5 90
26 47 34 66	32 56 23 54	24 41 18 68	16 35 21 76	14 19 16 92	7 8 13 103
27 55 21 58	23 47 23 63	16 35 24 74	18 40 16 71	9 21 11 90	8 14 12 97
23 49 27 64	24 47 29 63	18 43 22 66	26 38 19 73	17 23 11 88	11 18 11 93
25 53 21 60	23 52 19 58	27 47 17 62	17 31 24 80	13 17 19 94	11 18 15 93
18 49 19 64	25 48 21 62	13 37 16 72	20 38 20 73	17 23 5 88	15 22 14 89
22 50 24 63	22 44 18 66	15 40 20 69	18 38 14 73	9 11 24 100	13 21 7 90
23 52 22 61	17 40 19 70	27 45 12 64	20 34 14 77	19 26 15 85	9 15 15 96
31 51 22 62	15 42 25 68	20 30 30 79	18 28 19 83	13 22 6 89	13 21 12 90
21 42 26 71	23 52 14 58	15 40 20 69	11 29 25 82	9 15 12 96	13 20 12 91
22 47 20 66	12 43 20 67	20 45 11 64	25 43 13 68	13 18 15 93	9 19 10 94
19 45 19 68	27 51 19 59	24 36 19 73	19 31 16 80	9 20 12 91	17 20 11 93
20 45 25 68	21 43 25 67	17 31 25 78	16 28 22 83	18 23 12 88	13 14 6 99
26 50 26 63	30 47 16 63	17 39 14 70	21 34 23 77	12 17 14 94	6 7 6 106

(continued on next page)

Table B.6. *Daily rainfall at Durban (continued)*

4 7 12 106	5 9 7 106	13 15 10 95	25 36 13 74	19 51 24 60	19 44 28 66
12 15 11 98	7 11 5 104	8 12 12 98	7 24 14 86	24 56 19 55	24 53 21 57
6 14 13 99	6 9 7 106	12 16 15 94	14 31 12 79	18 51 21 60	32 50 18 60
17 21 6 92	7 10 6 105	10 19 18 91	18 29 17 81	21 54 20 57	15 36 29 73
5 10 14 103	8 9 14 106	18 27 9 83	11 28 21 82	24 53 18 58	17 51 26 59
15 19 12 94	11 15 11 100	11 18 14 92	21 38 23 72	19 47 24 64	33 60 24 50
9 16 7 97	12 15 7 100	11 21 12 89	15 40 14 70	28 52 20 59	24 51 23 59
9 14 11 99	6 10 16 105	10 22 11 88	16 39 16 71	13 44 23 67	19 50 24 60
11 16 9 97	12 20 14 95	15 23 16 87	17 39 16 71	27 54 24 56	26 55 21 55
7 14 9 99	16 22 11 93	16 24 6 86	21 38 18 72	18 51 19 59	19 50 22 60
14 16 5 97	10 17 8 98	6 14 18 96	12 35 15 75	22 52 26 58	22 53 22 57
4 7 9 106	6 15 6 100	13 26 11 84	19 38 20 72	19 56 21 54	30 53 22 57
6 12 14 101	9 15 6 100	10 24 18 86	20 39 21 71	25 58 25 52	17 45 22 65
12 20 8 93	8 12 15 103	18 32 9 78	23 41 25 70	20 58 17 52	27 50 20 60
14 16 12 97	13 19 15 96	10 23 12 87	24 43 19 68	26 55 18 55	20 43 26 67
12 14 13 99	16 21 11 94	15 25 11 85	21 38 20 73	11 47 22 63	24 49 26 61
10 15 7 98	11 16 5 99	11 21 14 89	17 37 22 74	26 58 21 52	16 51 21 59
8 12 2 101	8 10 16 105	10 24 11 85	21 42 10 69	22 53 12 57	23 56 18 54
6 6 12 107	10 18 8 97	12 25 13 85	13 31 17 80	18 43 25 67	21 51 27 59
10 12 9 101	12 16 6 99	13 26 20 84	14 35 21 76	26 50 26 60	27 57 18 53
8 11 14 102	6 10 13 105	18 33 13 77	19 42 16 69	21 50 27 60	22 48 15 62
12 17 8 96	15 17 8 93	22 28 15 82	18 39 21 72	18 56 22 54	22 41 27 69
8 13 11 100	6 10 10 100	13 21 18 89	18 42 26 69	28 60 22 50	19 46 25 64
13 16 7 97	9 14 8 96	14 26 15 84	23 50 19 61	26 54 19 56	22 52 25 58
8 10 12 103	9 13 9 97	14 27 15 83	22 46 20 65	25 47 21 63	26 55 20 55
8 14 6 99	6 13 12 97	12 28 17 82	21 44 20 67	16 43 24 67	25 49 17 61
5 12 11 103	10 19 13 91	22 33 18 77	19 43 24 68	21 51 23 59	20 41 22 69
13 18 6 97	14 22 9 88	16 29 16 81	21 48 18 63	25 53 22 57	18 43 27 67
7 11 11 104	13 17 5 93	18 29 16 81	17 45 21 66	25 50 19 60	27 52 17 58
9 15 12 100	3 9 10 101	14 27 20 83	20 49 18 62	15 44 25 66	25 42 23 68
14 18 7 97	10 16 14 94	18 33 18 77	14 47 21 64	26 54 23 56	20 40 25 70
9 11 10 104	14 20 8 90	17 33 18 77	28 54 16 57	17 51 26 59	21 45 19 65
5 12 13 103	7 14 10 96	12 34 19 76	21 42 30 69	25 60 19 50	26 43 29 67
12 20 9 95	13 17 11 93	22 41 19 69	22 51 21 60	27 54 24 56	29 46 23 64
15 17 7 98	12 15 12 95	20 38 18 72	23 50 24 61	21 51 14 59	

Cape Town homicides and suicides

Table B.7 contains, for each of the 313 weeks making up the years 1986–1991 inclusive, the numbers of deaths recorded at the Salt River state mortuary, Cape Town, as falling in each of the following five categories (in order): firearm homicide, nonfirearm homicide, firearm suicide, nonfirearm suicide, legal intervention homicide.

The layout of the table is as follows: first read down the first block of five columns, then similarly down the second to sixth blocks and the following page.

Dr Leonard Lerer is thanked for making these data available.

Table B.7. *Homicides and suicides, Cape Town, 1986–1991*

0 27 0 4 0	0 17 1 4 0	1 13 0 1 0	6 16 0 2 0	1 21 1 1 0	0 32 0 1 2
3 15 0 1 0	1 18 0 0 0	3 14 0 3 0	2 19 0 2 0	7 29 1 2 0	1 41 0 1 0
1 15 1 2 3	0 16 0 3 0	2 6 1 3 0	2 19 2 1 0	2 22 0 2 0	10 25 1 3 2
1 20 1 0 1	2 17 0 3 2	2 12 2 3 1	3 27 0 0 0	3 29 0 4 0	4 25 2 1 0
3 13 3 1 1	2 15 0 2 1	3 10 1 0 0	0 9 4 1 1	2 27 0 2 0	3 14 1 0 0
2 11 1 1 0	2 24 1 0 1	2 13 0 2 0	0 10 2 2 0	2 11 0 2 0	1 26 0 2 0
0 10 1 2 1	3 20 1 2 1	2 22 1 0 1	0 7 1 5 0	2 16 2 1 0	5 20 3 3 0
2 11 1 2 0	1 21 0 3 0	3 11 2 0 0	1 28 1 2 0	3 16 0 0 0	2 15 1 4 0
3 16 2 1 8	3 20 0 2 0	1 9 2 2 1	1 22 1 2 1	3 16 1 3 0	4 20 1 1 0
1 17 1 3 0	2 17 1 4 0	0 13 1 0 0	4 19 0 0 1	3 20 1 3 0	1 16 1 3 0
3 11 2 3 0	3 18 0 4 0	1 12 1 1 1	3 11 1 0 0	0 11 0 3 1	3 21 0 2 0
4 13 0 3 3	3 18 0 0 0	5 14 0 4 1	1 19 0 4 0	2 18 0 3 0	1 26 3 1 0
0 14 1 2 2	0 21 0 0 0	1 19 1 1 0	1 14 3 4 0	0 15 0 0 1	1 18 1 1 0
1 14 1 3 0	1 26 2 2 1	0 13 1 1 2	1 11 2 5 0	2 14 0 1 1	3 23 1 1 1
1 15 0 0 1	2 18 0 0 1	0 18 0 0 0	5 17 1 3 0	1 16 0 0 0	1 28 1 1 0
1 15 0 0 0	1 11 1 3 0	0 22 0 2 1	0 21 0 1 1	1 12 0 1 0	7 28 0 2 1
0 7 1 2 1	3 17 1 0 1	1 22 0 1 0	1 26 0 0 1	2 24 1 3 1	6 34 1 2 0
2 16 0 3 0	2 12 0 1 0	2 7 1 2 0	2 14 1 1 0	0 13 0 2 1	8 38 1 1 0
2 11 0 0 2	3 12 4 0 1	3 17 0 2 1	1 16 1 2 0	1 11 0 1 0	7 26 2 3 1
10 15 1 3 1	1 11 2 1 0	2 27 0 2 0	1 21 1 1 1	0 22 0 3 0	2 18 2 0 0
1 45 0 2 0	1 17 0 1 0	5 22 0 2 0	0 17 2 2 0	1 17 1 0 1	2 20 2 2 0
1 15 1 1 0	0 11 3 2 0	4 21 0 3 0	0 21 0 4 0	2 27 1 2 0	0 25 0 1 0
7 25 0 4 2	0 16 0 4 0	3 12 0 2 0	1 22 1 3 0	1 20 1 2 0	1 23 0 2 0
4 25 1 1 1	0 16 0 2 1	4 24 2 4 0	1 18 0 4 1	2 10 0 3 0	4 35 0 4 0
2 12 0 0 0	1 10 0 2 3	0 15 0 0 0	3 16 1 3 0	2 16 1 0 0	6 17 0 2 0
4 17 1 2 1	2 12 1 0 0	1 18 1 3 2	2 17 1 2 0	2 22 0 4 0	2 20 0 4 0
4 20 2 3 0	1 19 4 2 0	0 25 1 3 0	2 13 0 2 1	3 28 1 1 1	2 26 0 3 1
4 11 0 5 0	2 17 0 1 1	2 35 0 1 2	4 14 1 1 0	0 23 0 1 0	4 17 1 5 3
2 17 0 1 1	1 14 1 1 0	1 22 2 2 0	3 33 1 3 1	4 13 0 1 0	3 22 0 3 1
6 11 1 3 0	3 17 0 0 1	3 11 0 6 0	0 18 0 1 0	1 21 1 3 0	3 21 0 2 0
2 8 0 2 1	1 14 1 4 0	1 19 0 1 0	4 19 0 1 1	3 34 0 1 0	4 28 0 2 1
3 13 2 2 2	1 25 2 4 0	0 12 0 0 0	2 23 1 4 0	0 25 1 0 1	1 21 0 3 1
4 11 1 4 0	0 14 0 4 0	3 24 0 1 0	1 20 0 3 0	2 30 0 3 0	5 19 1 0 4
1 6 2 3 0	2 19 0 1 1	2 23 1 0 1	2 18 0 1 0	1 26 1 1 0	4 18 0 2 0
0 19 0 0 0	0 12 0 1 0	3 17 1 2 1	0 17 0 3 1	1 26 2 3 0	0 17 0 0 1
1 18 0 0 0	1 22 1 2 1	0 17 1 4 1	4 18 1 1 0	1 21 0 2 0	2 25 2 2 2
0 6 3 1 0	1 7 0 1 0	1 18 0 1 1	1 20 0 1 0	1 22 2 1 1	4 16 2 1 0
1 8 0 0 0	0 13 0 1 0	1 30 1 2 0	1 21 1 0 0	1 23 1 1 0	4 19 2 2 0

(continued on next page)

Table B.7. *Homicides and suicides (continued)*

0 19 1 3 4	1 33 1 0 0	3 31 0 3 0	3 35 2 1 3	2 25 0 4 1	8 17 1 1 0
2 24 1 0 0	1 29 2 1 0	0 52 1 3 2	2 22 1 6 1	5 27 1 0 0	14 29 2 4 0
1 19 1 3 0	2 27 1 4 3	4 27 0 2 1	1 18 1 4 3	2 22 0 1 3	9 22 0 2 0
3 24 0 3 0	4 27 0 0 2	2 21 1 4 1	3 12 2 1 1	7 25 1 2 0	5 23 0 1 0
0 16 3 3 2	4 29 1 4 0	6 17 1 2 0	1 10 0 0 2	3 21 0 6 0	5 27 2 1 0
2 18 0 2 2	2 14 2 2 0	1 19 0 3 0	0 19 1 2 1	4 17 0 1 1	15 41 0 2 0
4 23 0 2 0	2 31 1 1 0	4 16 0 2 0	2 19 0 0 0	7 17 1 1 0	7 29 0 3 0
5 20 1 4 1	9 22 2 3 0	3 12 1 3 0	0 25 0 1 0	7 30 1 1 0	11 35 3 2 0
3 20 1 1 0	6 21 4 3 9	5 15 1 2 0	4 18 1 2 2	12 13 0 3 1	11 25 0 3 0
0 14 0 1 2	3 26 0 0 0	4 25 0 2 1	6 26 0 2 0	12 23 0 2 1	4 41 0 3 1
5 18 0 6 0	2 18 0 4 0	1 20 0 4 2	7 18 1 2 1	14 19 1 3 2	
5 33 0 6 0	4 28 0 1 0	4 21 0 2 0	3 15 1 1 0	1 29 1 1 0	
4 26 0 4 4	1 22 0 4 0	1 27 1 0 1	2 25 0 2 2	12 32 2 2 0	
2 21 0 6 1	4 40 1 1 0	5 21 1 4 0	2 27 0 0 0	2 22 0 1 0	
1 16 0 1 0	3 31 1 4 1	4 20 0 0 2	9 25 1 1 0	5 28 0 2 1	

References

Adke, S.R. and Deshmukh, S.R. (1988). Limit distribution of a high order Markov chain. *J. R. Statist. Soc.* B **50**, 105–108.

Aitchison, J. (1986). *The Statistical Analysis of Compositional Data*. Chapman & Hall, London.

Albert, P.S. (1991). A two-state Markov mixture model for a time series of epileptic seizure counts. *Biometrics* **47**, 1371–1381.

Albert, P.S. (1994). A Markov model for sequences of ordinal data from a relapsing-remitting disease. *Biometrics* **50**, 51–60.

Albert, P.S., McFarland, H.F., Smith, M.E. and Frank, J.A. (1994). Time series for modelling counts from a relapsing-remitting disease: application to modelling disease activity in multiple sclerosis. *Statist. Med.* **13**, 453–466.

Al-Osh, M.A. and Aly, E.-E.A.A. (1992). First order autoregressive time series with negative binomial and geometric marginals. *Commun. Statist.-Theory Meth.* **21**, 2483–2492.

Al-Osh, M.A. and Alzaid, A.A. (1987). First-order integer-valued autoregressive (INAR(1)) process. *J. Time Ser. Anal.* **8**, 261–275.

Al-Osh, M.A. and Alzaid, A.A. (1988). Integer-valued moving average (INMA) process. *Statist. Papers* **29**, 281–300.

Al-Osh, M.A. and Alzaid, A.A. (1991). Binomial autoregressive moving average models. *Stoch. Models* **7**, 261–282.

Alzaid, A.A. and Al-Osh, M.A. (1988). First-order integer-valued autoregressive (INAR(1)) process: distributional and regression properties. *Statist. Neerl.* **42**, 53–61.

Alzaid, A.A. and Al-Osh, M.A. (1990). An integer-valued pth-order autoregressive structure (INAR(p)) process. *J. Appl. Prob.* **27**, 314–324.

Alzaid, A.A. and Al-Osh, M.A. (1993). Some autoregressive moving average processes with generalized Poisson marginal distributions. *Ann. Inst. Statist. Math.* **45**, 223–232.

Azzalini, A. (1982). Approximate filtering of parameter driven processes. *J. Time Ser. Anal.* **3**, 219–223.

Azzalini, A. (1983). Maximum likelihood estimation of order m for stationary stochastic processes. *Biometrika* **70**, 381–387.

Azzalini, A. (1994). Logistic regresssion for autocorrelated data with application to repeated measures. *Biometrika* **81**, 767–775.

Azzalini, A. and Bowman, A.W. (1990). A look at some data on the Old Faithful geyser. *Appl. Statist.* **39**, 357–365.

Baldi, P., Chauvin, Y., Hunkapiller, T. and McClure, M.A. (1994). Hidden Markov models of biological primary sequence information. *Proc. Nat. Acad. Sci. U.S.A.* **91**, 1059–1063.

Ball, F.G. and Rice, J.A. (1992). Stochastic models for ion channels: introduction and bibliography. *Math. Biosci.* **112**, 189–206.

Basawa, I.V. and Prakasa Rao, B.L.S. (1980). *Statistical Inference for Stochastic Processes.* Academic Press, London.

Baum, L.E. (1972). An inequality and associated maximization technique in statistical estimation for probabilistic functions of Markov processes. *Proc. Third Symposium on Inequalities*, ed. O. Shisha. Academic Press, New York, 1–8.

Baum, L.E. and Eagon, J.A. (1967). An inequality with applications to statistical estimation for probabilistic functions of Markov processes and to a model for ecology. *Bull. Amer. Math. Soc.* **73**, 360–363.

Baum, L.E. and Petrie, T. (1966). Statistical inference for probabilistic functions of finite state Markov chains. *Ann. Math. Statist.* **37**, 1554–1563.

Baum, L.E. and Sell, G.R. (1968). Growth transformations for functions on manifolds. *Pacific J. Math.* **27**, 211–227.

Baum, L.E., Petrie, T., Soules, G. and Weiss, N. (1970). A maximization technique occurring in the statistical analysis of probabilistic functions of Markov chains. *Ann. Math. Statist.* **41**, 164–171.

Bickel, P.J. and Ritov, Y. (1996). Inference in hidden Markov models I: Local asymptotic normality in the stationary case. *Bernoulli* **2**, 199–228.

Billingsley, P. (1961). Statistical methods in Markov chains. *Ann. Math. Statist.* **32**, 12–40.

Bisgaard, S. and Travis, L.E. (1991). Existence and uniqueness of the solution of the likelihood equations for binary Markov chains. *Statist. Prob. Letters* **12**, 29–35.

Blight, P.A. (1989). Time series formed from the superposition of discrete renewal processes. *J. Appl. Prob.* **26**, 189–195.

Block, H.W., Langberg, N.A. and Stoffer, D.S. (1988). Bivariate exponential and geometric autoregressive and autoregressive moving average models. *Adv. Appl. Prob.* **20**, 798–821.

Box, G.E.P. and Jenkins, G.M. (1976). *Time Series Analysis, Forecasting and Control*, revised edition. Holden-Day, Oakland, California.

Brockwell, P.J. and Davis, R.A. (1991). *Time Series: Theory and Methods*, second edition. Springer, New York.

Buishand, T.A. (1978). The binary DARMA(1,1) process as a model for wet-dry sequences. Technical note 78-01, Dept. of Mathematics,

REFERENCES

Statistics Division, Agricultural University, Wageningen, The Netherlands.

Campbell, M.J. (1994). Time series regression for counts: an investigation of the relationship between sudden infant death syndrome and environmental temperature. *J. R. Statist. Soc.* A **157**, 191–208.

Cane, V.R. (1978). On fitting low-order Markov chains to behaviour sequences. *Anim. Behav.* **26**, 332–338.

Chang, T.J., Delleur, J.W. and Kavvas, M.L. (1987). Application of discrete autoregressive moving average models for estimation of daily runoff. *J. Hydrol.* **91**, 119–135.

Chang, T.J., Kavvas, M.L. and Delleur, J.W. (1984a). Modeling of sequences of wet and dry days by binary discrete autoregressive moving average processes. *J. Clim. Appl. Meteorol.* **23**, 1367–1378.

Chang, T.J., Kavvas, M.L. and Delleur, J.W. (1984b). Daily precipitation modeling by discrete autoregressive moving average processes. *Water Resour. Res.* **20**, 565–580.

Chatfield, C. (1973). Statistical inference regarding Markov chain models. *Appl. Statist.* **22**, 7–20.

Chatfield, C. and Lemon, R.E. (1970). Analysing sequences of behavioural events. *J. Theoret. Biol.* **29**, 427-445.

Churchill, G.A. (1989). Stochastic models for heterogeneous DNA sequences. *Bull. Math. Biol.* **51**, 79–94.

Churchill, G.A. (1992). Hidden Markov chains and the analysis of genome structure. *Computers Chem.* **16**, 107–115.

Cox, D.R. (1981). Statistical analysis of time series: some recent developments. *Scand. J. Statist.* **8**, 93–115.

Cox, D.R. (1990). Role of models in statistical analysis. *Statist. Sci.* **5**, 169–174.

Cox, D.R. and Lewis, P.A.W. (1966). *The Statistical Analysis of Series of Events.* Methuen, London.

Cox, D.R. and Miller, H.D. (1965). *The Theory of Stochastic Processes.* Chapman & Hall, London.

Cox, D.R. and Snell, E.J. (1989). *Analysis of Binary Data*, second edition. Chapman & Hall, London.

Cramér, H. (1946). *Mathematical Methods of Statistics.* Princeton University Press, Princeton, New Jersey.

Delleur, J.W., Chang, T.J. and Kavvas, M.L. (1989). Simulation models of sequences of wet and dry days. *J. Irrig. and Drainage Engr.* **115**, 344–357.

Devore, J.L. (1976). A note on the estimation of parameters in a Bernoulli model with dependence. *Ann. Statist.* **4**, 990–992.

Diggle, P. and Westcott, M. (1985). Contribution to the discussion of Lawrance and Lewis (1985). *J. R. Statist. Soc.* B **47**, 192–193.

Dion, J.-P., Gauthier, G. and Latour, A. (1992). Branching processes with immigration and integer-valued time series. Preprint, Université

du Québec à Montréal.
Du Jin-Guan and Li Yuan (1991). The integer-valued autoregressive (INAR(p)) model. *J. Time Ser. Anal.* **12**, 129–142.
Efron, B., and Tibshirani, R.J. (1993). *An Introduction to the Bootstrap*. Chapman & Hall, New York.
Elliott, R.J., Aggoun, L. and Moore, J.B. (1995). *Hidden Markov Models: Estimation and Control*. Springer, New York.
Fahrmeir, L. and Kaufmann, H. (1987). Regression models for non-stationary categorical time series. *J. Time Ser. Anal.* **8**, 147–160.
Fildes, R. (1991). Review of Harvey (1989) and of West and Harrison (1989). *J. Opl. Res. Soc.* **42**, 1031–1033.
Franke, J. and Seligmann, T. (1993). Conditional maximum-likelihood estimates for INAR(1) processes and their application to modelling epileptic seizure counts. *Developments in Time Series Analysis*, ed. T. Subba Rao. Chapman & Hall, London, 310–330.
Fredkin, D.R. and Rice, J.A. (1992a). Bayesian restoration of single-channel patch clamp recordings. *Biometrics* **48**, 427–448.
Fredkin, D.R. and Rice, J.A. (1992b). Maximum likelihood estimation and identification directly from single-channel recordings. *Proc. R. Soc. London* B **249**, 125–132.
Gaver, D.P. and Lewis, P.A.W. (1980). First-order autoregressive gamma sequences and point processes. *Adv. Appl. Prob.* **12**, 727–745.
Gill, P.E., Murray, W., Saunders, M.A. and Wright, M.H. (1986). User's Guide for NPSOL: a Fortran package for nonlinear programming. Report SOL 86-2, Department of Operations Research, Stanford University.
Grimmett, G.R. and Stirzaker, D.R. (1992). *Probability and Random Processes*, second edition. Oxford University Press, Oxford.
Grunwald, G.K., Raftery, A.E. and Guttorp, P. (1993). Time series of continuous proportions. *J. R. Statist. Soc.* B **55**, 103-116.
Guttorp, P. (1986). On binary time series obtained from continuous time point processes describing rainfall. *Water Resour. Res.* **22**, 897–904.
Guttorp, P. (1995). *Stochastic Modeling of Scientific Data*. Chapman & Hall, London.
Guttorp, P., Newton, M.A. and Abkowitz, J.L. (1990). A stochastic model for haematopoiesis in cats. *IMA J. of Math. Appl. in Med. and Biol.* **7**, 125–143.
Guttorp, P. and Thompson, M.L. (1990). Nonparametric estimation of intensities for sampled counting processes. *J. R. Statist. Soc.* B **52**, 157–173.
Haines, L.M., Munoz, W.P. and van Gelderen, C.J. (1989). ARIMA modelling of birth data. *J. Appl. Statist.* **16**, 55–67.
Harvey, A.C. (1989). *Forecasting, Structural Time Series Models and the Kalman Filter*. Cambridge University Press, Cambridge.
Harvey, A.C. and Fernandes, C. (1989a). Time series models for count

or qualitative observations. *J. Bus. Econ. Statist.* **7**, 407–422.

Harvey, A.C. and Fernandes, C. (1989b). Time series models for insurance claims. *J. Inst. Actuaries* **116**, 513–528.

Holden, R.T. (1987). Time series analysis of a contagious process. *J. Amer. Statist. Ass.* **82**, 1019–1026.

Hopkins, A., Davies, P. and Dobson, C. (1985). Mathematical models of patterns of seizures: their use in the evaluation of drugs. *Arch. Neurol.* **42**, 463–467.

Hughes, J.P. (1993). A class of stochastic models for relating synoptic atmospheric patterns to local hydrologic phenomena. Ph.D. dissertation, Dept. of Statistics, University of Washington.

Jacobs, P.A. and Lewis, P.A.W. (1978a). Discrete time series generated by mixtures I: Correlational and runs properties. *J. R. Statist. Soc.* B **40**, 94–105.

Jacobs, P.A. and Lewis, P.A.W. (1978b). Discrete time series generated by mixtures II: Asymptotic properties. *J. R. Statist. Soc.* B **40**, 222–228.

Jacobs, P.A. and Lewis, P.A.W. (1978c). Discrete time series generated by mixtures III: Autoregressive processes $(DAR(p))$. Technical report NPS55-78-022, Naval Postgraduate School, Monterey, California.

Jacobs, P.A. and Lewis, P.A.W. (1983). Stationary discrete autoregressive-moving average time series generated by mixtures. *J. Time Ser. Anal.* **4**, 19–36.

Jamshidian, M. and Jennrich, R.I. (1993). Conjugate gradient acceleration of the EM algorithm. *J. Amer. Statist. Ass.* **88**, 221–228.

Juang, B.H. (1985). Maximum-likelihood estimation for mixture multivariate stochastic observations of Markov chains. *AT&T Tech. J.* **64**, 1235–1249.

Juang, B.H. and Rabiner, L.R. (1991). Hidden Markov models for speech recognition. *Technometrics* **33**, 251–272.

Kanter, M. (1975). Autoregression for discrete processes mod 2. *J. Appl. Prob.* **12**, 371–375.

Kashiwagi, N. and Yanagimoto, T. (1992). Smoothing serial count data through a state-space model. *Biometrics* **48**, 1187–1194.

Katz, R.W. (1981). On some criteria for estimating the order of a Markov chain. *Technometrics* **23**, 243–249.

Kaufmann, H. (1987). Regression models for nonstationary categorical time series: asymptotic estimation theory. *Ann. Statist.* **15**, 79–98.

Kedem, B. (1976). Sufficient statistics associated with a two-state second-order Markov chain. *Biometrika* **63**, 127–132.

Kedem, B. (1980). *Binary Time Series.* Marcel Dekker, New York.

Keenan, D.M. (1982). A time series analysis of binary data. *J. Amer. Statist. Ass.* **77**, 816–821.

Kelly, F.P. (1979). *Reversibility and Stochastic Networks.* Wiley, Chichester.

Kemeny, J.G., Snell, J.L. and Knapp, A.W. (1976). *Denumerable Markov Chains.* Springer, New York.

Klimko, L.A. and Nelson, P.I. (1978). On conditional least squares estimation for stochastic processes. *Ann. Statist.* **6**, 629–642.

Klotz, J. (1973). Statistical inference in Bernoulli trials with dependence. *Ann. Statist.* **1**, 373–379.

Krogh, A., Brown, M., Mian, I.S., Sjölander, K. and Haussler, D. (1994). Hidden Markov models in computational biology: applications to protein modeling. *J. Mol. Biol.* **235**, 1501–1531.

Langberg, N.A. and Stoffer, D.S. (1987). Moving-average models with bivariate exponential and geometric distributions. *J. Appl. Prob.* **24**, 48–61.

Lawrance, A.J. (1976). On conditional and partial correlation. *Amer. Statistician* **30**, 146–149.

Lawrance, A.J. (1979). Partial and multiple correlation for time series. *Amer. Statistician* **33**, 127–130.

Lawrance, A.J. (1982). The innovation distribution of a gamma distributed autoregressive process. *Scand. J. Statist.* **9**, 234–236.

Lawrance, A.J. and Lewis, P.A.W. (1980). The exponential autoregressive-moving average EARMA(p, q) process. *J. R. Statist. Soc.* B **42**, 150–161.

Lawrance, A.J. and Lewis, P.A.W. (1981). A new autoregressive time series model in exponential variables (NEAR(1)). *Adv. Appl. Prob.* **13**, 826–845.

Lawrance, A.J. and Lewis, P.A.W. (1985). Modelling and residual analysis of nonlinear autoregressive time series in exponential variables. *J. R. Statist. Soc.* B **47**, 165–183.

Le, N.D., Leroux, B.G. and Puterman, M.L. (1992). Reader reaction: Exact likelihood evaluation in a Markov mixture model for time series of seizure counts. *Biometrics* **48**, 317–323.

Le Cam, L. and Yang, G.L. (1990). *Asymptotics in Statistics. Some Basic Concepts.* Springer, New York.

Leroux, B.G. (1992). Maximum-likelihood estimation for hidden Markov models. *Stoch. Processes Appl.* **40**, 127–143.

Leroux, B.G. and Puterman, M.L. (1992). Maximum-penalized-likelihood estimation for independent and Markov-dependent mixture models. *Biometrics* **48**, 545–558.

Levinson, S.E., Rabiner, L.R. and Sondhi, M.M. (1983). An introduction to the application of the theory of probabilistic functions of a Markov process to automatic speech recognition. *Bell System Tech. J.* **62**, 1035–1074.

Lewis, P.A.W. (1985). Some simple models for continuous variate time series. *Water Resour. Bull.* **21**, 635–644.

Lewis, P.A.W., McKenzie, E. and Hugus, D.K. (1989). Gamma processes. *Stoch. Models* **5**, 1–30.

REFERENCES

Li, W.K. (1991). Testing model adequacy for some Markov regression models for time series. *Biometrika* **78**, 83–89.

Li, W.K. (1994). Time series models based on generalized linear models: some further results. *Biometrics* **50**, 506–511.

Li, W.K. and Kwok, M.C.O. (1990). Some results on the estimation of a higher order Markov chain. *Commun. Statist.-Simul. Comp.* **19**, 363–380.

Liang, K.-Y. and Zeger, S.L. (1989). A class of logistic regression models for multivariate binary time series. *J. Amer. Statist. Ass.* **84**, 447–451.

Linhart, H. and Zucchini, W. (1986). *Model Selection.* Wiley, New York.

Little, R.J.A. and Rubin, D.B. (1987). *Statistical Analysis with Missing Data.* Wiley, New York.

Ljung, G.M. (1993). On outlier detection in time series. *J. R. Statist. Soc.* B **55**, 559–567.

Lloyd, E.H. (1980). *Handbook of Applicable Mathematics, Vol. 2: Probability.* Wiley, New York.

McCullagh, P. and Nelder, J.A. (1989). *Generalized Linear Models*, second edition. Chapman & Hall, London.

MacDonald, I.L. and Lerer, L.B. (1994). A time series analysis of trends in firearm-related homicide and suicide. *Int. J. Epidemiol.* **23**, 66–72.

McInnes, F. and Jack, M. (1988). Automatic speech recognition using word reference patterns. *Speech Technology: A Survey*, eds M. Jack and J. Laver. Edinburgh University Press, Edinburgh, 1–68.

McKenzie, E. (1981). Extending the correlation structure of exponential autoregressive-moving-average processes. *J. Appl. Prob.* **18**, 181–189.

McKenzie, E. (1985a). Contribution to the discussion of Lawrance and Lewis (1985). *J. R. Statist. Soc.* B **47**, 187–188.

McKenzie, E. (1985b). Some simple models for discrete variate time series. *Water Resour. Bull.* **21**, 645–650.

McKenzie, E. (1986). Autoregressive moving-average processes with negative-binomial and geometric marginal distributions. *Adv. Appl. Prob.* **18**, 679–705.

McKenzie, E. (1987). Innovation distributions for gamma and negative binomial autoregressions. *Scand. J. Statist.* **14**, 79–85.

McKenzie, E. (1988a). The distributional structure of finite moving-average processes. *J. Appl. Prob.* **25**, 313–321.

McKenzie, E. (1988b). Some ARMA models for dependent sequences of Poisson counts. *Adv. Appl. Prob.* **20**, 822–835.

Mehran, F. (1989). Analysis of discrete longitudinal data: infinite-lag Markov models. *Statistical Data Analysis and Inference*, ed. Y. Dodge. Elsevier Science Publishers, Amsterdam, 533–541.

Meilijson, I. (1989). A fast improvement to the EM algorithm on its own terms. *J. R. Statist. Soc.* B **51**, 127–138.

Morgan, B.J.T. (1976). Markov properties of sequences of behaviours. *Appl. Statist.* **25**, 31–36.

Munoz, W.P., Haines, L.M. and van Gelderen, C.J. (1987). An analysis of the maternity data of Edendale Hospital in Natal for the period 1970–1985. Part 1: Trends and seasonality. Internal report, Edendale Hospital.

Noble, B. (1969). *Applied Linear Algebra*. Prentice Hall, Englewood Cliffs, New Jersey.

Numerical Algorithms Group (1992). *NAG Fortran Library (Mark 15)*. Numerical Algorithms Group, Oxford.

Pegram, G.G.S. (1980). An autoregressive model for multilag Markov chains. *J. Appl. Prob.* **17**, 350–362.

Poritz, A.B. (1988). Hidden Markov models: a guided tour. *Proc. 1988 Int. Conf. Acoust., Speech, Signal Processing*. IEEE Press, New York, 7–13.

Press, W.H., Flannery, B.P., Teukolsky, S.A. and Vetterling, W.T. (1986). *Numerical Recipes: The Art of Scientific Computing*. Cambridge University Press, Cambridge.

Priestley, M.B. (1981). *Spectral Analysis and Time Series*. Academic Press, London.

Raftery, A.E. (1985a). A model for high-order Markov chains. *J. R. Statist. Soc.* B **47**, 528–539.

Raftery, A.E. (1985b). A new model for discrete-valued time series: autocorrelations and extensions. *Rassegna di Metodi Statistici ed Applicazioni*, **3–4**, 149–162.

Raftery, A.E. and Tavaré, S. (1994). Estimation and modelling repeated patterns in high order Markov chains with the mixture transition distribution model. *Appl. Statist.* **43**, 179–199.

Reeves, G.K. (1993). Goodness-of-fit tests in two-state processes. *Biometrika* **80**, 431–442.

Rugg, D.J. and Buech, R.R. (1990). Analyzing time budgets with Markov chains. *Biometrics* **46**, 1123–1131.

Rydén, T. (1994). Consistent and asymptotically normal parameter estimates for hidden Markov models. *Ann. Statist.* **22**, 1884–1895.

Rydén, T. (1995). Estimating the order of hidden Markov models. *Statistics* **26**, 345–354.

Schimert, J. (1992). A high order hidden Markov model. Ph.D. dissertation, University of Washington.

Schwarz, G. (1978). Estimating the dimension of a model. *Ann. Statist.* **6**, 461–464.

Seneta, E.E. (1981). *Non-negative Matrices and Markov Chains*. Springer, New York.

Singh, A.C. and Roberts, G.R. (1992). State space modelling of cross-classified time series of counts. *Int. Statist. Rev.* **60**, 321–335.

Steutel, F.W. and van Harn, K. (1979). Discrete analogues of self-decomposability and stability. *Ann. Prob.* **7**, 893–899.

Stoffer, D.S. (1985). Central limit theorems for finite Walsh–Fourier transforms of weakly stationary time series. *J. Time Ser. Anal.* **6**, 261–267.

Stoffer, D.S. (1987). Walsh–Fourier analysis of discrete-valued time series. *J. Time Ser. Anal.* **8**, 449–467.

Stoffer, D.S. (1990). Multivariate Walsh–Fourier analysis. *J. Time Ser. Anal.* **11**, 57–73.

Stoffer, D.S. (1991). Walsh–Fourier analysis and its statistical applications. *J. Amer. Statist. Ass.* **86**, 461–485.

Stuart, A. and Ord, J.K. (1991). *Kendall's Advanced Theory of Statistics, Vol. 2*, fifth edition. Edward Arnold, London.

Teicher, H. (1954). On the multivariate Poisson distribution. *Skand. Aktuarietidskr.* **37**, 1–9.

Thompson, E.A. (1983). Optimal sampling for pedigree analysis: parameter estimation and genotypic uncertainty. *Theor. Pop. Biol.* **24**, 39–58.

Titterington, D.M. (1984). Comments on 'Application of the conditional population-mixture model to image segmentation'. *IEEE Trans. Pattern Anal. Machine Intell.* **6**, 656–658.

Titterington, D.M. (1990). Some recent research in the analysis of mixture distributions. *Statistics* **21**, 619–641.

West, M. and Harrison, P.J. (1989). *Bayesian Forecasting and Dynamic Models*. Springer, New York.

West, M., Harrison, P.J. and Migon, H.S. (1985). Dynamic generalized linear models and Bayesian forecasting. *J. Amer. Statist. Ass.* **80**, 73–97.

Whittaker, J. (1990). *Graphical Models in Applied Multivariate Statistics*. Wiley, Chichester.

Wichmann, B.A. and Hill, I.D. (1982). An efficient and portable pseudorandom number generator. *Appl. Statist.* **31**, 188-190. Correction, *Appl. Statist.* **33** (1984), 123.

Winkler, R.L. (1989). Contribution to the discussion of Harvey and Fernandes (1989a). *J. Bus. Econ. Statist.* **7**, 419–422.

Woolhiser, D.A. (1992). Modeling daily precipitation — progress and problems. *Statistics in the Environmental and Earth Sciences*, eds A.T. Walden and P. Guttorp. Edward Arnold, London, 71–89.

Zeger, S.L. (1988). A regression model for time series of counts. *Biometrika* **75**, 621–629.

Zeger, S.L. and Liang, K.-Y. (1991). Feedback models for discrete and continuous time series. *Statistica Sinica* **1**, 51–64.

Zeger, S.L. and Qaqish, B. (1988). Markov regression models for time series: a quasi-likelihood approach. *Biometrics* **44**, 1019–1031.

Zucchini, W. and Adamson, P.T. (1984). The occurrence and severity of drought in South Africa. Water Research Commission report no. 92/1/84, Water Research Commission, Pretoria, South Africa.

Zucchini, W., Adamson, P.T. and McNeill, L. (1992). A model for Southern African rainfall. *S. Afr. J. Sci.* **88**, 103–109.

Zucchini, W. and Guttorp, P. (1991). A hidden Markov model for space-time precipitation. *Water Resour. Res.* **27**, 1917–1923.

Author index

Abkowitz, J.L., 56, 220
Adamson, P.T., 184, 187, 225
Adke, S.R., 15, 217
Aggoun, L., 56, 220
Aitchison, J., 116, 132, 217
Albert, P.S., 10, 41, 56, 66, 80, 97, 146, 217
Al-Osh, M.A., 21–32, 49, 217
Aly, E.-E.A.A., 21, 217
Alzaid, A.A., 21–32, 49, 217
Azzalini, A., 4, 11–12, 16, 43, 78, 138–140, 145, 207, 217, 218

Baldi, P., 56, 218
Ball, F.G., 56, 218
Basawa, I.V., 6, 218
Baum, L.E., 58, 59, 63, 95, 218
Bickel, P.J., 95, 218
Billingsley, P., 6, 218
Bisgaard, S., 10, 48, 139, 218
Blight, P.A., 48, 218
Block, H.W., 32–37, 218
Bowman, A.W., 4, 16, 138–140, 145, 207, 218
Box, G.E.P., 20, 139, 218
Brockwell, P.J., 9, 14, 218
Brown, M., 56, 222
Buech, R.R., 162, 224
Buishand, T.A., 17, 20, 218

Campbell, M.J., 45, 219
Cane, V.R., 162, 219
Chang, T.J., 17, 20, 219
Chatfield, C., 162, 219

Chauvin, Y., 56, 218
Churchill, G.A., 56, 219
Cox, D.R., 7, 37, 42, 50, 69, 71, 72, 76, 107, 155, 219
Cramér, H., 9, 219

Davies, P., 146, 221
Davis, R.A., 9, 14, 218
Delleur, J.W., 17, 20, 219
Deshmukh, S.R., 15, 217
Devore, J.L., 48, 219
Diggle, P., 50, 219
Dion, J.-P., 21, 219
Dobson, C., 146, 221
Du Jin-Guan, 21, 31, 220

Eagon, J.A., 58, 218
Efron, B., 97, 220
Elliott, R.J., 56, 220

Fahrmeir, L., 40–41, 119, 220
Fernandes, C., 45, 50, 220, 221
Fildes, R., 45, 220
Flannery, B.P., 90, 137, 157, 224
Frank, J.A., 41, 56, 217
Franke, J., 21, 146, 220
Fredkin, D.R., 56, 64, 94, 220

Gauthier, G., 21, 219
Gaver, D.P., 24, 43, 220
Gill, P.E., 91, 220
Grimmett, G.R., 6, 220
Grunwald, G.K., 116, 220

Guttorp, P., 4, 56, 64, 78, 84, 107, 116, 121, 122, 179, 190, 220, 226

Haines, L.M., 152–159, 220, 224
Harrison, P.J., 5, 45–47, 50, 220, 225
Harvey, A.C., 5, 45–46, 50, 220, 221
Haussler, D., 56, 222
Hill, I.D., 143, 225
Holden, R.T., 41, 221
Hopkins, A., 146, 221
Hughes, J.P., 130–131, 221
Hugus, D.K., 46, 222
Hunkapiller, T., 56, 218

Jack, M., 68, 223
Jacobs, P.A., 17–20, 36, 47, 221
Jamshidian, M., 94, 221
Jenkins, G.M., 20, 139, 218
Jennrich, R.I., 94, 221
Juang, B.H., 55, 56, 59, 64, 221

Kanter, M., 47, 221
Kashiwagi, N., 47, 221
Katz, R.W., 105, 177, 221
Kaufmann, H., 40–41, 119, 220, 221
Kavvas, M.L., 17, 20, 219
Kedem, B., 43, 48, 177–178, 221
Keenan, D.M., 42, 48, 221
Kelly, F.P., 102, 221
Kemeny, J.G., 56, 222
Klimko, L.A., 26, 222
Klotz, J., 47, 48, 222
Knapp, A.W., 56, 222
Krogh, A., 56, 222
Kwok, M.C.O., 13, 16, 223

Langberg, N.A., 32–37, 218, 222
Latour, A., 21, 219
Lawrance, A.J., 9, 22–24, 222

Le, N.D., 56, 80, 146–147, 151, 208, 222
Le Cam, L., 95, 222
Lemon, R.E., 162, 219
Lerer, L.B., 198, 223
Leroux, B.G., 56, 57, 66, 80, 84, 91–95, 106, 129, 132, 136, 146–147, 151, 208, 222
Levinson, S.E., 55, 61–64, 78, 80, 92, 222
Lewis, P.A.W., 17–20, 22–24, 36, 43, 46, 47, 69, 71, 72, 76, 219–222
Li, W.K., 13, 16, 40, 223
Li Yuan, 21, 31, 220
Liang, K.-Y., 41, 223, 225
Linhart, H., 188, 223
Little, R.J.A., 64, 223
Ljung, G.M., 97, 223
Lloyd, E.H., 13, 223

McClure, M.A., 56, 218
McCullagh, P., 155, 187, 223
MacDonald, I.L., 198, 223
McFarland, H.F., 41, 56, 217
McInnes, F., 68, 223
McKenzie, E., 21–30, 32, 36, 46, 47, 222, 223
McNeill, L., 184, 225
Mehran, F., 16, 223
Meilijson, I., 94, 223
Mian, I.S., 56, 222
Migon, H.S., 45–47, 225
Miller, H.D., 7, 219
Moore, J.B., 56, 220
Morgan, B.J.T., 162, 223
Munoz, W.P., 152–159, 220, 224
Murray, W., 91, 220

Nelder, J.A., 155, 187, 223
Nelson, P.I., 26, 222
Neuwirth, L.P., 56
Newton, M.A., 56, 220
Noble, B., 7, 224

AUTHOR INDEX

Numerical Algorithms Group, 91, 137, 207, 224

Ord, J.K., 9, 225

Pegram, G.G.S., 4, 13, 18, 111, 112, 144, 224
Petrie, T., 58, 59, 63, 95, 218
Poritz, A.B., 56, 224
Prakasa Rao, B.L.S., 6, 218
Press, W.H., 90, 137, 157, 224
Priestley, M.B., 35, 224
Puterman, M.L., 56, 57, 66, 80, 84, 91–94, 106, 129, 130, 132, 136, 146–147, 151, 208, 222

Qaqish, B., 12, 37–41, 50, 202, 225

Rabiner, L.R., 55, 59, 61–64, 78, 80, 92, 221, 222
Raftery, A.E., 4, 13–16, 111, 112, 116, 136, 144, 202, 220, 224
Reeves, G.K., 202, 224
Rice, J.A., 56, 64, 94, 218, 220
Ritov, Y., 95, 218
Roberts, G.R., 47, 224
Rubin, D.B., 64, 223
Rugg, D.J., 162, 224
Rydén, T., 95, 106, 224

Saunders, M.A., 91, 220
Schimert, J., 14, 115, 224
Schwarz, G., 15, 106, 224
Seligmann, T., 21, 146, 220
Sell, G.R., 58, 218
Seneta, E.E., 7, 224
Singh, A.C., 47, 224
Sjölander, K., 56, 222
Smith, M.E., 41, 56, 217
Snell, E.J., 50, 219
Snell, J.L., 56, 222
Sondhi, M.M., 55, 61–64, 78, 80, 92, 222

Soules, G., 58, 59, 63, 218
Steutel, F.W., 29, 224
Stirzaker, D.R., 6, 220
Stoffer, D.S., 32–37, 49, 119, 218, 222, 224, 225
Stuart, A., 9, 225

Tavaré, S., 14, 15, 224
Teicher, H., 27, 125, 225
Teukolsky, S.A., 90, 137, 157, 224
Thompson, E.A., 56, 225
Thompson, M.L., 4, 220
Tibshirani, R.J., 97, 220
Titterington, D.M., 106, 225
Travis, L.E., 10, 48, 139, 218

van Gelderen, C.J., 152–159, 220, 224
van Harn, K., 29, 224
Vetterling, W.T., 90, 137, 157, 224

Weiss, N., 58, 59, 63, 218
West, M., 5, 45–47, 50, 220, 225
Westcott, M., 50, 219
Whittaker, J., 67, 225
Wichmann, B.A., 143, 225
Winkler, R.L., 50, 225
Woolhiser, D.A., 184, 225
Wright, M.H., 91, 220

Yanagimoto, T., 47, 221
Yang, G.L., 95, 222

Zeger, S.L., 12, 37–41, 43–45, 50, 202, 223, 225
Zucchini, W., 56, 78, 84, 107, 121, 122, 179, 184, 187, 188, 190, 223, 225, 226

Subject index

ACF and PACF
  BGARMA models, 34–35
  binomial ARMA models, 29–30
  binomial-hidden Markov
    model, 74–77
  DARMA models, 18–20
  epileptic seizure series, 148
  geometric AR(1) model, 22
  INAR(1) model, 31
  INMA(q) model, 32
  loglinear regression models, 44
  Markov chains, 8–10
  negative binomial AR(1)
    model, 25
  Old Faithful geyser series,
    139–142
  Poisson ARMA models, 25,
    27–28
  Poisson-hidden Markov models,
    70–74, 77
  Raftery models, 15
  thinly traded shares series,
    179–181
Advertising
  application of dynamic
    generalized linear models, 47
AIC and BIC, 105–106, 137
  epileptic seizure series, 147
  homicide series, 193–198
  Koeberg wind direction series,
    173–176
  locusts series, 164–166
  Markov chains, 106
  Old Faithful geyser series,
    144–145
Akaike's information criterion, *see
  also* AIC and BIC
  Durban rainfall series, 188
Animal behaviour, 162–168
Applications, 137–202
ARMA models
  based on renewal processes, 48
  bivariate geometric, *see*
    BGARMA models
  discrete, *see* DARMA models
  Gaussian, 17, 152–153, 159
  geometric marginals, 21–23, 36
  integer-valued, *see* INAR and
    INMA models
  negative binomial marginals,
    23–25
  Poisson marginals, 25–28
Autocorrelation function, *see*
  ACF and PACF

Baum–Welch algorithm, 91–95
  computational considerations,
    64
  in speech processing, 58–64
  proofs of results, 203–206
Bayesian information criterion,
  see also AIC and BIC
  Edendale births series, 155,
    158, 161
  evapotranspiration series, 177
  Raftery models, 15
  thinly traded shares series, 179

SUBJECT INDEX

Bayesian models, *see* State-space models
Bernoulli series, *see* Binary series; Binomial series
BGARMA models, 32–37
Binary hidden Markov model
  evapotranspiration series, 177–178
  Old Faithful geyser series, 140–146
  runlength distribution, 86–90
  thinly traded shares series, 180
Binary series, *see also* Binomial series
  Durban rainfall series, 184
  evapotranspiration series, 177
  Kanter's autoregressive model, 47
  Markov regression models, 37
  multivariate, 56
  nonhomogeneous Markov chain models for, 11–12
  Old Faithful geyser series, 138
  parameter-driven models, 42–43
  thinly traded shares, 178
  with covariates, 12, 38
Binomial-hidden Markov model, 66
  ACF and PACF, 74–77
  distributional properties, 82–84
  Edendale births series, 157–159
  homicide series, 192, 195
  locusts series, 166
  multivariate, 122
  reversibility, 102–105
  with change-point, 193
  with Poisson numbers of trials, 133
Binomial series
  Markov regression models, 37–38
  models based on thinning, 29–30
Bird-song patterns
  application of Raftery models, 15
Bonferroni bounds, 100
  epileptic seizure series, 149
Bootstrap, 96–97, 143, 182

Categorical series, 40, 119–121
  homicide series, 200
  Koeberg wind direction series, 168–176
Clipped data
  methods for, 48
Compositional data, 116
Conditional independence, 67
  contemporaneous, 121–125
Conditional likelihood
  Markov chains, 10
  Markov regression model, 38
Coronary risk factors
  model for, 12
Covariates
  categorical series, 40
  hidden Markov models, 128–130
  loglinear regression models, 44
  Markov regression models, 37–42
Cross-correlations
  multinomial-hidden Markov models, 117–119
  multivariate hidden Markov models, 122–125

DARMA models, 17–20
Decoding
  global, 64
  localized, 64
DNA sequences
  application of Raftery models, 15
Durban rainfall series, 184–191
  data, 213–214
Dynamic generalized linear models, 46–47, 50

SUBJECT INDEX

Edendale births series, 152–162
  data, 208–209
EM algorithm, 80, 92–94
  Baum–Welch algorithm, 58–64
Empirical model, 107
Epileptic seizure series, 21, 56, 146–152
  data, 208
Evapotranspiration series, 177–178
Exponential family
  Markov regression model, 39
  state-space models, 46

Forecast distribution
  dynamic generalized linear model, 47
  from joint hidden Markov models, 134
  hidden Markov models, 82
  Keenan's model, 43
  Old Faithful geyser series, 146
Forecast function
  structural models, 46
Forecast p-values and pseudo-residuals
  definition, 99
  epileptic seizure series, 150
  homicide series, 199
Forward and backward probabilities
  definition, 59
  pseudo-residuals calculated from, 99
Forward-backward algorithm, 58–60, 79

Generalized linear models
  dynamic, 46–47
  Markov regression models, 37
Geometric series
  bivariate, see BGARMA models
  models based on thinning, 21–23

Hidden Markov models
  asymptotic properties of maximum likelihood estimators, 95
  categorical series, 119–121, 168–176
  computation of likelihood, 79–80
  conditional independence, 67
  continuous-valued series, 68
  correlation properties, 69
  covariates, 128–130
  definition and notation, 65–69
  distributional properties, 80–90
  estimation software, 137, 207
  forecast distribution, 82
  joint models for trials and successes, 133–135
  Markov chain homogeneous but not stationary, 129–130
  Markov chain nonhomogeneous, 130–133, 170–172, 188–191
  missing data, 81
  most likely states, 64–65
  multinomial, 115–121, 168–176, 200–201
  multivariate, 121–128, 163–166, 180
  parameter estimation, 90–95
  parametric bootstrap, 96–97, 143, 182
  second-order, 110–115, 142
  speech processing, 55–65
  state-dependent probabilities, 66
  use of optimization software, 91
Higher-order Markov chains, 12–16, see also Pegram models; Raftery models
  evapotranspiration series, 177
  Old Faithful geyser series, 138–140
  parameter estimation, 15–16
  second-order, 110

234   SUBJECT INDEX

Homicides and suicides, 191–201
   data, 214–216

INAR and INMA models, 28, 31–32
Interpersonal relationships
   application of Raftery models, 15
Ion channel modelling, 56

Koeberg wind direction series, 168–176, 207

Labour statistics
   application of generalized Raftery models, 16
Likelihood, *see also* Quasi-likelihood
   categorical series, 120
   conditional, *see* Conditional likelihood
   forward-backward algorithm, 58–60
   hidden Markov models, 77–80, *see also* Baum–Welch algorithm
   multinomial-hidden Markov models, 116
   multivariate hidden Markov models, 122
   nonhomogeneous hidden Markov models, 131
   nonhomogeneous Markov chain, 12
   saturated Markov chain, 10
   second-order hidden Markov models, 113–115
   split data, 95, 106
Locusts series, 162–168
   data, 209–211
Logistic autoregressive models, *see* Markov regression models
Logistic-linear models, 37
Logit transformation, 12

Durban rainfall series, 186–189
Edendale births series, 154
Markov regression model, 37
Loglinear regression models, 43–45

Markov chains, 6–16
   ACF, PACF, 8–10
   as parameter process, 55
   conditional maximum likelihood estimation, 10
   diagonalizable, 7
   Durban rainfall series, 185
   eigenvalues and eigenvectors, 7–8, 87
   estimating the order of, 105
   evapotranspiration series, 177–178
   higher order, *see* Higher-order Markov chains
   homogeneous, 7, 129
   irreducible, 7
   Koeberg wind direction series, 173–176
   locusts series, 162–168
   nonhomogeneous, 11–12, 185
   Old Faithful geyser series, 138
   runlength for binary chain, 86–90
   saturated, 6–11
   stationary distribution, 7
   transition probability matrix, 7
Markov regression models, 37–42, 50
   Edendale births series, 154–162
   estimation software, 155
   linear contagion model, 41
   multivariate, 41
   parameter estimation, 38
Missing data
   binary series, 12
Mixture transition distribution models, *see* Raftery models
Multinomial-hidden Markov models, 115–121

SUBJECT INDEX

homicide series, 200–201
Multiple sclerosis modelling, 56
Multivariate hidden Markov models, 121–128
  mixed discrete–continuous series, 126

NDARMA models, *see* DARMA models
Negative binomial series
  models based on thinning, 23–25
Nelder–Mead algorithm, 90, 137, 157

Observation-driven models, 50
  categorical series, 119
  Markov regression models, 37
Occupational mobility
  application of Raftery models, 15
Old Faithful geyser series, 138–146
  data, 207
Outliers
  binary series, 101
  epileptic seizure series, 149–151
  hidden Markov models, 97
Overdispersion
  in series of counts, 55
  loglinear regression models, 44
  of Poisson-hidden Markov model, 71
  use of negative binomial distribution, 68

PACF, 9, 77, *see also* ACF and PACF
Parameter-driven models, 42–45, 50
Parameter process, 55
Partial autocorrelation function, 9, 77, *see also* ACF and PACF
Pegram models, 13–14, 111–113

Old Faithful geyser series, 144
Poisson-hidden Markov model, 66
  ACF, 70–74, 148
  distributional properties, 83–84
  Edendale births series, 159
  epileptic seizure series, 146–152
  homicides series, 194
  multivariate, 122
  parametric bootstrap, 96–97
  reversibility, 102–104
  with trend, 194
Poisson series, *see also* Markov regression models
  linear contagion model, 41
  models based on thinning, 25–28
  parameter-driven models, 43
  state-space models, 45
Precipitation amounts
  application of DARMA models, 20
Pseudo-residuals
  definition, 98
  epileptic seizure series, 149
p-values, 98

Quasi-likelihood
  loglinear regression models, 44
  Markov regression models, 38

Raftery models, 13–16, 111–113
  Koeberg wind direction series, 175–176
  Old Faithful geyser series, 144
Rainfall wet–dry day series, *see also* Durban rainfall series
  application of DARMA models, 20
  multivariate, 56, 121
Renewal processes
  models based on, 48
Reversibility, 101
  binomial ARMA(1,1), 30
  DMA(1) model, 19
  hidden Markov models, 109

Poisson AR and MA models, 25, 27
Runlengths
  binary hidden Markov model, 86–90
Runoff
  application of DARMA models, 20

Sales forecasting
  application of dynamic generalized linear models, 47
Seasonality
  Durban rainfall series, 185–191
  Edendale births series, 154–162
  hidden Markov models, 128, 133
  Koeberg wind direction series, 170
  Markov regression models, 37
  nonhomogeneous hidden Markov models, 131
Speech processing, 55–65
State-dependent probabilities, 66
State-space models, 45–47, 50
Structural models, 45–46
Substantive model, 107

Telephone calls
  application of dynamic generalized linear models, 47
Thinly traded shares series, 178–183
  data, 211–212
Thinning models, 21–32
  binomial marginals, 29–30
  geometric marginals, 21–23
  negative binomial marginals, 23–25
  Poisson marginals, 25–28
Trend
  Edendale births series, 152–162
  hidden Markov models, 128, 133
  homicide series, 191–201

locusts series, 162–168
Markov regression models, 37
piecewise constant, 193–201

Viterbi algorithm, 65

Walsh–Fourier analysis, 49
Wind direction
  application of Raftery models, 15
  at Koeberg, 168–176
Wind power
  application of Raftery models, 15

Yule–Walker equations
  DARMA models, 18
  Raftery models, 14